Singular Perturbation Theory

Lindsay A. Skinner

Singular Perturbation Theory

Lindsay A. Skinner
Department of Mathematical Sciences
University of Wisconsin - Milwaukee
Milwaukee, Wisconsin 53201
USA
skinner4@cox.net

ISBN 978-1-4419-9957-3   e-ISBN 978-1-4419-9958-0
DOI 10.1007/978-1-4419-9958-0
Springer New York Dordrecht Heidelberg London

Library of Congress Control Number: 2011928077

Mathematical Subject Classification 2010: 34E05, 34E10, 34E15, 34E20

Printed on acid-free paper

Springer is part of Springer Science+Business Media (www.springer.com)

Preface

There are many fine books on singular perturbation theory and how to solve singular perturbation problems. Many readers of this book will already be familiar with one or more of them. What distinguishes this book from the others is its rigorous development and rigorous application of the method of matched asymptotic expansions. The point of view is that certain functions have a certain structure for which this method is valid and these are precisely the kinds of functions that arise in a wide variety of differential equation and integration problems.

This book is intended to serve primarily as a supplement or follow-up to a typical first year graduate course in asymptotic and perturbation analysis. Hopefully it will also prove to be a valuable companion to all those who do, or wish to do, rigorous work in the field. The basic theory for the book is presented in Chapter 1. Then there are four chapters in which this theory is applied to a sequence of ordinary differential equation problems. There are a number of previously unpublished results. One of these is the uniformly valid expansion at the end of Chapter 4 for a problem involving logarithms once studied by L. E. Fraenkel. Another is the unexpectedly simple uniformly valid Bessel function expansion established at the end of Chapter 3. All the differential equations chosen for study in the text are linear, but this is not because of any limitation of the theory. Indeed, much has been done, and much more can be done, in applying the theory to nonlinear problems, as noted in Exercise 3.2, for example.

Another unique feature of this book is its inclusion of several Maple programs for computing the terms of the various asymptotic expansions that arise in solving the problems. Developing these was a nice break from the otherwise sometimes tedious hard analysis in the book. They could well have just been designated as exercises, indeed a few have been, but at the same time, until one gets familiar with symbolic programming, there is nothing like having a few model programs around to show the way, or at least a possible way. Studying the programs will also reveal some mathematical manipulations that are not fully developed within the text itself.

There is a short bibliography of key references at the end of the book. My proof that the solution to Problem C is uniformly of order one, and the use of this fact to complete the solution to Problem C, is essentially due to R. E. O'Malley, Jr., as described in Section 5.4 of [12], the book by D. R. Smith, although a different approach to this problem, one comparable to mine for Problems D, E and I, is used by O'Malley in [7]. Also, the inclusion of Problem D, which serves as a nice transition from Problem C to Problem E, comes about as a result of my having first seen it discussed in [12]. Problem G is the one previously studied by L. E. Fraenkel. From his work in Part II of [3] I saw how to prove my asymptotic results for this problem by first establishing a convergent series solution. Similarly, I first became aware of the intriguing Problem I by reading J. Kevorkian's account of it in the 1981 version of [5]. My own work on it began in [11]. My analysis for Problem B is an outgrowth of results presented in [8].

All of this started for me with the publication of the first edition of [13], the wonderful little book by Milton Van Dyke, in which the idea that matching was somehow just a counting game and not an idea dependent on notions of overlapping domains of validity first became apparent. L. E. Fraenkel's three paper series, which came soon after, was the first to try to show that the reason this counting game works is a consequence of function structure. The present book is an outgrowth of my attempts to clarify this connection and to build on it.

A recurring theme in the book is that for each problem we first establish the form of a uniformly valid asymptotic expansion for its solution. Then, knowing this form, we are able to proceed to calculate the appropriate inner and outer expansions from the differential equation for the problem, and thus determine the terms of the uniformly valid expansion. Of course, when a problem is new, one is likely to proceed differently. I discovered the form of the uniformly valid asymptotic expansion for the solution to Problem G, for example, by first calculating a few terms of its inner expansion, just a power series if the initial conditions are treated as parameters, and then examining the outer expansions of these terms. Once I had this uniformly valid form, then from it the form of the outer expansion of the solution, together with the required matching conditions for it, was readily determined. There was no need for any kind of special matching considerations due to the logarithms. In determing uniformly valid expansions for the solutions to Problem H and Problem I, an important bit of insight, also a key step in [9], was to factor out the oscillatory parts, as they do not have matching inner and outer expansions.

At the end of each chapter there are exercises to do. Some of them demonstrate results in the chapter, some fill in missing steps in the chapter, and some are included to prepare for the next chapter. Of course, everyone is encouraged to work all the exercises and, indeed, to add their own. Readers interested in just a basic understanding, however, need only read Chapter 1, possibly work Exercises 1.2 and 1.3, then read Section 2.1 and work Exercise

2.1. Finally they should read Section 3.1, maybe work Exercise 3.1, and get some experience with the Maple program for Problem C.

Lindsay A. Skinner
San Diego, California

Contents

Chapter 1
Uniform Expansion Theory

1.1 Introduction

Roughly speaking, a function $z(x, \varepsilon)$ is a singular perturbation of $z(x, 0)$ if $z(x, 0)$ fails to approximate $z(x, \varepsilon)$ for all x of interest when ε is small. Uniformly valid approximations for such functions can often be found by the so-called method of matched asymptotic expansions. The purpose of this book is to present a rigorous development of this method and its application to integral and differential equation problems.

Suppose, for example, that $z(x, \varepsilon) = 1/(x + 2x^3 + \varepsilon)$ and suppose we wish to estimate

$$F(\varepsilon) = \int_0^1 z(x, \varepsilon)\, dx \qquad (1.1)$$

for $0 < \varepsilon \ll 1$. Away from $x = 0$, if ε is small enough, we can neglect it compared to x and x^3, and approximate $z(x, \varepsilon)$ by $z(x, 0) = 1/(x + 2x^3)$. Near $x = 0$, on the other hand, x^3 is negligible compared to x, so $z(x, \varepsilon)$ is approximately $1/(x + \varepsilon)$. Furthermore, away from $x = 0$, $1/(x + \varepsilon)$ is approximately $1/x$, and near $x = 0$, $1/(x + 2x^3)$ is approximately $1/x$. In a mouthful, the near $x = 0$ approximation of the away from $x = 0$ approximation of $z(x, \varepsilon)$ matches the away from $x = 0$ approximation of the near $x = 0$ approximation of $z(x, \varepsilon)$. An important consequence of this apparent coincidence is that the composite function

$$c(x, \varepsilon) = \frac{1}{x + \varepsilon} + \frac{1}{x + 2x^3} - \frac{1}{x} \qquad (1.2)$$

has the same approximations near $x = 0$ and away from $x = 0$ as $z(x, \varepsilon)$. In fact we are about to prove the precise statement,

$$z(x, \varepsilon) = c(x, \varepsilon) + O(\varepsilon) \qquad (1.3)$$

1

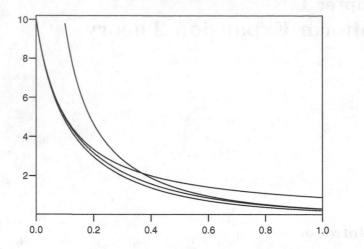

Fig. 1.1 Graph of $z(x, \varepsilon)$ together with the approximations $1/(x + \varepsilon)$, $1/(x + 2x^3)$ and $c(x, \varepsilon)$ when $\varepsilon = 0.1$.

uniformly as $\varepsilon \to 0^+$ for $0 \leq x \leq 1$. That is, there exists positive numbers ε_o and B such that $|z(x, \varepsilon) - c(x, \varepsilon)| \leq B\varepsilon$ for all $(x, \varepsilon) \in [0, 1] \times (0, \varepsilon_o]$.

From the integral of $c(x, \varepsilon)$, and the fact that $\ln(1 + \varepsilon) = O(\varepsilon)$ as $\varepsilon \to 0^+$, it follows from (1.3) that

$$F(\varepsilon) = \ln(1/\varepsilon) - (1/2)\ln 3 + O(\varepsilon) \tag{1.4}$$

as $\varepsilon \to 0^+$. Figure 1.1 shows the approximations $1/(x + \varepsilon)$ and $1/(x + 2x^2)$, along with $z(x, \varepsilon)$ and $c(x, \varepsilon)$, for $\varepsilon = 0.1$.

1.2 Uniform Expansion Theorem

As is customary, we will say $f(x) \in C^N([0, 1])$ if $f(x)$ is differentiable up to N times on the open interval $(0, 1)$ and each derivative extends continuously to $[0, 1]$. If $f(x) \in C^N([0, 1])$ for all $N \geq 0$, we say $f(x) \in C^\infty([0, 1])$. A basic asymptotic result is that if $f(x) \in C^N([0, 1])$, then

$$f(x) = \sum_{n=0}^{N-1} f^{[n]}(0)x^n + O(x^N) \tag{1.5}$$

on the interval $[0, 1]$, where $f^{[n]}(x) = \frac{1}{n!}\left(\frac{d}{dx}\right)^n f(x)$. Indeed, $f(x) \in C^N([0, 1])$ implies

$$\phi(x) = x^{-N}\left[f(x) - \sum_{n=0}^{N-1} f^{[n]}(0)x^n\right] \tag{1.6}$$

is in $C^0([0,1])$, and therefore is bounded on $[0,1]$. In particular, $\phi(0) = f^{[N]}(0)$. We will also say $f(x) \in C^N([1,\infty])$ if $\hat{f}(x) \in C^N([0,1])$, where $\hat{f}(x) = f(1/x)$. When this is the case, (1.5) holds on $[0,1]$ with $\hat{f}$ in place of f, and if we substitute $x = 1/X$, we get

$$f(X) = \sum_{n=0}^{N-1} \hat{f}^{[n]}(0)X^{-n} + O(X^{-N}) \tag{1.7}$$

on $[1,\infty]$. Moreover, $\hat{f}^{[1]}(x) = (-1/x^2)f^{[1]}(1/x) = -X^2\frac{d}{dX}f(X)$ and therefore $\hat{f}^{[n]}(0) = f^{[-n]}(\infty)$, where $f^{[-n]}(X) = \frac{1}{n!}\left(-X^2\frac{d}{dX}\right)^n f(X)$. This notation readily extends to functions of more than one variable. In particular, for $m, n \geq 0$,

$$f^{[m,-n]}(x,X) = \frac{1}{m!n!}\left(\frac{\partial}{\partial x}\right)^m\left(-X^2\frac{\partial}{\partial X}\right)^n f(x,X) \tag{1.8}$$

and we say $f(x,X) \in C^N([0,1] \times [0,\infty])$ if both $f(x,X)$ and $\hat{f}(x,X)$ are in $C^N([0,1] \times [0,1])$, where $\hat{f}(x,X) = f(x,1/X)$.

We are going to be dealing with functions $y(x,\varepsilon)$ that are, for some $\varepsilon_o > 0$, expressible in the form $y(x,\varepsilon) = f(x,x/\varepsilon)$ for all $(x,\varepsilon) \in [0,1] \times (0,\varepsilon_o]$, where $f(x,X) \in C^\infty([0,1] \times [0,\infty])$. If $y(x,\varepsilon)$ is one of these functions, then for any $\delta \in (0,1)$ and any $N \geq 1$,

$$y(x,\varepsilon) = O_N y(x,\varepsilon) + O(\varepsilon^N) \tag{1.9}$$

uniformly as $\varepsilon \to 0^+$ for $\delta \leq x \leq 1$, where

$$O_N y(x,\varepsilon) = \sum_{n=0}^{N-1} (\varepsilon/x)^n f^{[0,-n]}(x,\infty). \tag{1.10}$$

Indeed, there exists $B_N, \Delta_N > 0$ such that

$$\left|f(x,X) - \sum_{n=0}^{N-1} X^{-n} f^{[0,-n]}(x,\infty)\right| \leq B_N X^{-N} \tag{1.11}$$

whenever $(x,X) \in [0,1] \times [\Delta_N,\infty]$, and therefore

$$\left|f(x,x/\varepsilon) - \sum_{n=0}^{N-1} (\varepsilon/x)^n f^{[0,-n]}(x,\infty)\right| \leq C_N \varepsilon^N \tag{1.12}$$

for all $(x, \varepsilon) \in [\delta, 1] \times (0, d_N]$, where $C_N = B_N/\delta^N$ and $d_N = \delta/\Delta_N$. The function $O_N y(x, \varepsilon)$ is called the N-term outer expansion of $y(x, \varepsilon)$. Of course, $x^{-n} f^{[0,-n]}(x, \infty) = y^{[0,n]}(x, 0)$.

In addition to (1.9), for any $\varepsilon \in (0, \varepsilon_o]$ and integer $M \geq 1$,

$$y(x, \varepsilon) = I_M y(x, \varepsilon) + O(x^M) \tag{1.13}$$

as $x \to 0^+$, where

$$I_M y(x, \varepsilon) = \sum_{m=0}^{M-1} x^m f^{[m,0]}(0, x/\varepsilon). \tag{1.14}$$

We call $I_M y(x, \varepsilon)$ the M-term inner expansion of $y(x, \varepsilon)$. Note that it is the first M terms in the power series expansion of $Y(X, \varepsilon) = y(\varepsilon X, \varepsilon)$ as $\varepsilon \to 0^+$, assuming $X > 0$, followed by the substitution $X = x/\varepsilon$. Applying this idea to $O_N y(x, \varepsilon)$ leads to

$$I_M O_N y(x, \varepsilon) = \sum_{m=0}^{M-1} x^m \sum_{n=0}^{N-1} (\varepsilon/x)^n f^{[m,-n]}(0, \infty). \tag{1.15}$$

Similarly, the first N terms in the expansion of $I_M y(x, \varepsilon)$ as $\varepsilon \to 0^+$, assuming $x > 0$, yields

$$O_N I_M y(x, \varepsilon) = \sum_{n=0}^{N-1} (\varepsilon/x)^n \sum_{m=0}^{M-1} x^m f^{[m,-n]}(0, \infty). \tag{1.16}$$

Obviously, $I_M O_N y(x, \varepsilon) = O_N I_M y(x, \varepsilon)$ for any $M, N \geq 1$. Note also that $O_N I_M O_N y(x, \varepsilon) = I_M O_N y(x, \varepsilon)$ and $I_M O_N I_M y(x, \varepsilon) = O_N I_M y(x, \varepsilon)$. Therefore the N-term composite function $C_N y(x, \varepsilon) = [O_N + I_N - O_N I_N] y(x, \varepsilon)$ has the same M-term inner and outer expansions, for any $M \leq N$, as $y(x, \varepsilon)$. In addition, note that $O_N I_N y(x, \varepsilon)$ can be divided into a part which is finite at $x = 0$ and a part which equals 0 at $x = \infty$. The first of these parts is

$$\sum_{n=0}^{N-1} (\varepsilon/x)^n \sum_{m=n}^{N-1} x^m f^{[m,-n]}(0, \infty) = \sum_{m=0}^{N-1} \varepsilon^m (x/\varepsilon)^m \sum_{n=0}^{m} (\varepsilon/x)^n f^{[m,-n]}(0, \infty). \tag{1.17}$$

Thus it is clear that

$$C_N y(x, \varepsilon) = \sum_{n=0}^{N-1} \varepsilon^n [u_n(x) + v_n(x/\varepsilon)], \tag{1.18}$$

where

$$u_n(x) = x^{-n} \left[f^{[0,-n]}(x, \infty) - \sum_{m=0}^{n-1} x^m f^{[m,-n]}(0, \infty) \right], \tag{1.19}$$

$$v_m(X) = X^m \left[f^{[m,0]}(0, X) - \sum_{n=0}^{m} X^{-n} f^{[m,-n]}(0, \infty) \right]. \tag{1.20}$$

Also, $f(x, X) \in C^\infty([0, 1] \times [0, \infty])$ implies $u_n(x) \in C^\infty([0, 1])$ and $v_m(X) \in C^\infty([0, \infty])$. We can now state the theorem that is the basis for all our work in this text.

Theorem 1. *If for some $\varepsilon_o > 0$, $y(x, \varepsilon) = f(x, x/\varepsilon)$ for all $(x, \varepsilon) \in [0, 1] \times (0, \varepsilon_o]$, where $f(x, X) \in C^{2N}([0, 1] \times [0, \infty])$, then*

$$y(x, \varepsilon) = \sum_{n=0}^{N-1} \varepsilon^n [u_n(x) + v_n(x/\varepsilon)] + O(\varepsilon^N) \tag{1.21}$$

uniformly as $\varepsilon \to 0^+$ for $0 \le x \le 1$, where $u_n(x)$ and $v_n(X)$ are the functions defined by (1.19) and (1.20).

1.3 Some Calculations

Before turning to the proof of Theorem 1, let us see how it applies to the problem of Section 1.1. If we let $y(x, \varepsilon) = \varepsilon z(x, \varepsilon)$, then $y(x, \varepsilon) = f(x, x/\varepsilon)$, where $f(x, X) = 1/[1 + X(1 + 2x^2)]$, and obviously $f(x, X) \in C^\infty([0, 1] \times [0, \infty])$. From

$$y(x, \varepsilon) = \frac{\varepsilon/(x + 2x^3)}{1 + \varepsilon/(x + 2x^3)} \tag{1.22}$$

for $x \ne 0$, it is apparent that

$$O_N y(x, \varepsilon) = \sum_{n=1}^{N-1} (-1)^{n+1} \frac{\varepsilon^n}{(x + 2x^3)^n}. \tag{1.23}$$

In particular,

$$O_1 y(x, \varepsilon) = 0, \qquad O_2 y(x, \varepsilon) = \frac{\varepsilon}{x + 2x^3}. \tag{1.24}$$

Similarly,

$$y(\varepsilon X, \varepsilon) = \frac{1/(1 + X)}{1 + 2\varepsilon^2 X^3/(1 + X)}, \tag{1.25}$$

so we have

$$I_1 y(x, \varepsilon) = I_2 y(x, \varepsilon) = \frac{1}{1 + x/\varepsilon}. \tag{1.26}$$

Next, from either (1.24) or (1.26) we get

$$O_1 I_1 y(x, \varepsilon) = 0, \quad O_2 I_2 y(x, \varepsilon) = \varepsilon/x. \tag{1.27}$$

Therefore $C_1 y(x, \varepsilon) = 1/(1 + x/\varepsilon)$, or $u_0(x) = 0$, $v_0(X) = 1/(1 + X)$. Also

$$[C_2 - C_1]y(x, \varepsilon) = \frac{\varepsilon}{x + 2x^3} - \frac{\varepsilon}{x}, \tag{1.28}$$

and therefore $u_1(x) = -2x/(1 + 2x^2)$, $v_1(X) = 0$. Thus we see by Theorem 1 with $N = 2$ that

$$y(x, \varepsilon) = \frac{1}{1 + x/\varepsilon} - \frac{2\varepsilon x}{1 + 2x^2} + O(\varepsilon^2), \tag{1.29}$$

and this confirms (1.3).

1.4 The Proof

The validity of Theorem 1 is a direct consequence of the following more fundamental result.

Lemma 1. *If $f(x, X) \in C^0([0,1] \times [0, \infty])$, then $f(x, x/\varepsilon) = \phi(x, x/\varepsilon) + o(1)$, where $\phi(x, X) = f(x, \infty) + f(0, X) - f(0, \infty)$, uniformly as $\varepsilon \to 0^+$ for $0 \le x \le 1$.*

Proof. Let $\eta > 0$ be given. Then there exists $\delta > 0$ such that $|f(x, X) - f(0, X)| < \eta$ whenever $(x, X) \in [0, \delta] \times [0, \infty]$. Therefore

$$|f(x, x/\varepsilon) - f(0, x/\varepsilon)| < \eta, \qquad |f(x, \infty) - f(0, \infty)| < \eta \tag{1.30}$$

for all $(x, \varepsilon) \in [0, \delta] \times (0, \infty]$. Similarly, there exists $\Delta > 0$ such that $|f(x, X) - f(x, \infty)| < \eta$ whenever $(x, X) \in [0, 1] \times [\Delta, \infty]$, and therefore

$$|f(x, x/\varepsilon) - f(x, \infty)| < \eta, \qquad |f(0, x/\varepsilon) - f(0, \infty)| < \eta \tag{1.31}$$

for all $(x, \varepsilon) \in [\delta, 1] \times (0, d]$, where $d = \delta/\Delta$. Together, (1.30) and (1.31) show that $|f(x, x/\varepsilon) - \phi(x, x/\varepsilon)| < 2\eta$ whenever $(x, \varepsilon) \in [0, 1] \times (0, d]$.

Proof of Theorem 1. To prove Theorem 1, we will show that $f(x, X) \in C^{2N}([0, 1] \times [0, \infty])$ implies the slightly stronger statement

$$f(x, x/\varepsilon) = \sum_{n=0}^{N} \varepsilon^n [u_n(x) + v_n(x/\varepsilon)] + o(\varepsilon^N) \tag{1.32}$$

uniformly as $\varepsilon \to 0^+$ for $0 \le x \le 1$. By the lemma, we know this is true for $N = 0$. Assume it is true for $N = M > 0$ and assume $f(x, X) \in C^{2(M+1)}([0, 1] \times [0, \infty])$. Then $g(x, X) = X[f(x, X) - f(x, \infty)]$ is in $C^{2M+1}([0, 1] \times [0, \infty])$, and hence $h(x, X) \in C^{2M}([0, 1] \times [0, \infty])$, where $h(x, X) = x^{-1}[g(x, X) - g(0, X)]$. Therefore

$$h(x, x/\varepsilon) = \sum_{n=0}^{M} \varepsilon^n [\hat{u}_n(x) + \hat{v}_n(x/\varepsilon)] + o(\varepsilon^M) \qquad (1.33)$$

uniformly as $\varepsilon \to 0^+$ for $0 \le x \le 1$, where

$$\hat{u}_n(x) = x^{-n} \Big[h^{[0,n]}(x, \infty) - \sum_{m=0}^{n-1} x^m h^{[m,-n]}(0, \infty) \Big], \qquad (1.34)$$

$$\hat{v}_m(X) = X^m \Big[h^{[m,0]}(0, X) - \sum_{n=0}^{m} X^{-n} h^{[m,-n]}(0, \infty) \Big], \qquad (1.35)$$

in accordance with (1.19) and (1.20). But for $\varepsilon > 0$,

$$h(x, x/\varepsilon) = \varepsilon^{-1}[f(x, x/\varepsilon) - \phi(x, x/\varepsilon)], \qquad (1.36)$$

where $\phi(x, X)$ is the function defined in Lemma 1. Therefore, substituting into (1.33),

$$f(x, x/\varepsilon) = \phi(x, x/\varepsilon) + \varepsilon \sum_{n=0}^{M} \varepsilon^n [\hat{u}_n(x) + \hat{v}_n(x/\varepsilon)] + o(\varepsilon^{M+1}). \qquad (1.37)$$

Finally, note that $g^{[0,-n]}(x, \infty) = f^{[0,-n-1]}(x, \infty)$ and hence

$$h^{[0,-n]}(x, \infty) = x^{-1}[f^{[0,-n-1]}(x, \infty) - f^{[0,-n-1]}(0, \infty)]. \qquad (1.38)$$

Similarly,

$$h^{[m,0]}(0, X) = X[f^{[m+1,0]}(0, X) - f^{[m+1,0]}(0, \infty)]. \qquad (1.39)$$

Thus we see $\hat{u}_n(x) = u_{n+1}(x)$ and $\hat{v}_m(X) = v_{m+1}(X)$, and therefore (1.37) is the same as (1.32) with $N = M + 1$.

1.5 Computer Calculation

Calculating more than a few terms in the uniform expansion (1.21) for most any $y(x, \varepsilon) = f(x, x/\varepsilon)$ quickly becomes tedious work. One of our objectives in this book is to encourage the use of computer algebra software to do calculations like this for us. We will be doing this using the well-known product called Maple. As our first example, starting on a blank Maple worksheet, for the function $y(x, \varepsilon) = \varepsilon/(\varepsilon + x + 2x^3)$ of Section 1.3, by entering the sequence of commands

$$y := \varepsilon \cdot (\varepsilon + x + 2x^3)^{-1}; \quad N := 4;$$

$series(y, \varepsilon = 0, N)$;
$ONy := convert(\%, polynom)$;
$series(subs(x = \varepsilon \cdot X, y), \varepsilon = 0, N)$;
$INy := convert(\%, polynom)$;
$series(subs(x = \varepsilon \cdot X, ONy), \varepsilon = 0, N)$;
$INONy := convert(\%, polynom)$;
$series(subs(X = \varepsilon^{-1} \cdot x, INy), \varepsilon = 0, N + 1)$;
$ONINy := convert(\%, polynom)$;

we readily determine $O_4 y(x, \varepsilon)$, $I_4 y(x, \varepsilon)$ and two forms of $O_4 I_4 y(x, \varepsilon) = I_4 O_4 y(x, \varepsilon)$. Notice though that in the *series* command to determine $ONINy$ with $N = 4$ we had to ask for $N + 1$ terms. This is because of the way *series* deals with removable singularities. Other values of N may require making other adjustments, as may other functions, $y(x, \varepsilon)$.

Continuing on the same worksheet, if we now enter

for n **from** 0 **to** $N - 1$ **do**
$expand(x^n \cdot coeff(ONINy, \varepsilon, n))$;
$series(\%, x = 0, n)$;
$coeff(ONy, \varepsilon, n) - x^{-n} \cdot convert(\%, polynom)$;
$u_n := simplify(\%)$;
end do;

we get, in addition to $u_0(x) = 0$ and $u_1(x) = -2x/(1 + 2x^2)$ as in Section 1.3, the new results

$$u_2(x) = \frac{4(1 + x^2)}{(1 + 2x^2)^2}, \quad u_3(x) = \frac{8x(3 + 8x^2 + 6x^4)}{(1 + 2x^2)^3}. \quad (1.40)$$

The idea in this second step is to capture the sum in (1.19) from the coefficient of ε^n in $O_N I_N y(x, \varepsilon)$ as it appears in (1.16). We used Maple's *expand* command to convert x^n times this coefficient into a polynomial and thus avoid another removable singularity problem. By similar reasoning, looking at (1.15) and the sum in (1.20), if we add

for m **from** 0 **to** $N - 1$ **do**
$expand(X^{-m} \cdot coeff(INONy, \varepsilon, m))$;
$series(\%, X = \infty, m + 1)$;
$coeff(INy, \varepsilon, m) - X^m \cdot convert(\%, polynom)$;
$v_m := simplify(\%)$;
end do;

we immediately get

$$v_0(X) = \frac{1}{1 + X}, \quad v_1(X) = 0, \quad v_2(X) = -\frac{2(3X + 2)}{(1 + X)^2}, \quad v_3(X) = 0. \quad (1.41)$$

As for the original problem of approximating the function $F(\varepsilon)$ given by (1.1), concluding this computation with

$CNy := sum(\varepsilon^k \cdot (u_k + v_k), k = 0, N - 1);$
$int(subs(X = \varepsilon^{-1} \cdot x, CNy), x = 0..1)$ assuming $\varepsilon > 0$;
$F := series(\varepsilon^{-1} * \%, \varepsilon = 0, N);$

yields

$$F(\varepsilon) = \ln(1/\varepsilon) - (1/2)\ln 3 + a\varepsilon - 6\varepsilon^2 \ln(1/\varepsilon) + b\varepsilon^2 + O(\varepsilon^3), \qquad (1.42)$$

where $a = \frac{4}{3} + \frac{3\sqrt{2}}{2}\arctan\sqrt{2}$ and $b = \frac{59}{18} + 3\ln 3$. Note that we had to ask for 4 terms in the *series* command for $F(\varepsilon)$ in order not to miss a possible $\varepsilon^3 \ln(1/\varepsilon)$ contribution.

1.6 Two Corollaries

Before turning to differential equation problems we have a pair of corollaries to add to Theorem 1. It often happens that $y(x, \varepsilon) = f(x, x/\varepsilon, \varepsilon)$ for $0 \le x \le 1, 0 < \varepsilon \le \varepsilon_o$, where $f(x, X, \varepsilon) \in C^\infty([0, 1] \times [0, \infty] \times [0, \varepsilon_o])$. Then, of course,

$$f(x, X, \varepsilon) = \sum_{n=0}^{N-1} \varepsilon^n f^{[0,0,n]}(x, X, 0) + O(\varepsilon^N) \qquad (1.43)$$

uniformly as $\varepsilon \to 0^+$ for all $(x, X) \in [0, 1] \times [0, \infty]$, and therefore

$$y(x, \varepsilon) = \sum_{n=0}^{N-1} \varepsilon^n f^{[0,0,n]}(x, x/\varepsilon, 0) + O(\varepsilon^N) \qquad (1.44)$$

uniformly for $0 \le x \le 1$. Furthermore, we can apply Theorem 1 to each $f^{[0,0,n]}(x, x/\varepsilon, 0)$.

Corollary 1. *If* $y(x, \varepsilon) = f(x, x/\varepsilon, \varepsilon)$ *for all* $(x, \varepsilon) \in [0, 1] \times (0, \varepsilon_o]$, *where* $f(x, X, \varepsilon) \in C^\infty([0, 1] \times [0, \infty] \times [0, \varepsilon_o])$ *and* $\varepsilon_o > 0$, *then for any* $N \ge 0$,

$$y(x, \varepsilon) = \sum_{n=0}^{N-1} \varepsilon^n [u_n(x) + v_n(x/\varepsilon)] + O(\varepsilon^N) \qquad (1.45)$$

uniformly for $0 \le x \le 1$ *as* $\varepsilon \to 0^+$, *where* $u_n(x) \in C^\infty([0, 1])$, $v_n(X) \in C^\infty([0, \infty])$, $v_n(\infty) = 0$, *and, as in Theorem 1, the sum in (1.45) is the combination* $C_N y(x, \varepsilon) = [O_N + I_N - O_N I_N] y(x, \varepsilon)$ *of the outer and inner expansions*

$$O_N y(x, \varepsilon) = \sum_{n=0}^{N-1} \varepsilon^n y_n(x), \tag{1.46}$$

where $y_n(x) = y^{[0,n]}(x, 0)$, and

$$I_N y(x, \varepsilon) = \sum_{n=0}^{N-1} \varepsilon^n Y_n(x/\varepsilon), \tag{1.47}$$

where, $Y_n(X) = Y^{[0,n]}(X, 0)$, $Y(X, \varepsilon) = y(\varepsilon X, \varepsilon)$. Also, for any $M, N \geq 1$, $O_N I_M y(x, \varepsilon) = I_M O_N y(x, \varepsilon)$.

Proof. From (1.44) and Theorem 1, it is easy to see (1.45) holds with

$$u_k(x) = \sum_{n=0}^{k} x^{-n} \left(g_{k-n}^{[0,-n]}(x, \infty) - \sum_{m=0}^{n-1} g_{k-n}^{[m,-n]}(0, \infty) x^m \right), \tag{1.48}$$

$$v_k(X) = \sum_{m=0}^{k} X^m \left(g_{k-m}^{[m,0]}(0, X) - \sum_{n=0}^{m} g_{k-m}^{[m,-n]}(0, \infty) X^{-n} \right), \tag{1.49}$$

where $g_k(x, X) = f^{[0,0,k]}(x, X, 0)$, and it is clear that $u_k(x) \in C^\infty([0, 1])$, $v_k(X) \in C^\infty([0, \infty])$ and $v_k(\infty) = 0$. For the (independent) determination of these functions by outer and inner expansions, we defer to Exercise 1.3 at the end of the chapter.

It often happens too that we have $y(x, \varepsilon) = f(x, x/\varepsilon, \varepsilon)$ and $f(x, X, \varepsilon) = o(X^{-\infty})$ as $X \to \infty$. That is, $f^{[0,-n,0]}(x, \infty, \varepsilon) = 0$ for all $n \geq 0$, which means $O_N y(x, \varepsilon) = 0$ for all $N \geq 1$. Also, additional parameters may be present and of course we are not restricted to the interval $0 \leq x \leq 1$. We will need the following, for example, in which x is the additional parameter, in the next chapter.

Corollary 2. *If $y(x, t, \varepsilon) = f(x, t, t/\varepsilon, \varepsilon)$, where $f(x, t, T, \varepsilon) \in C^\infty([0, x_o] \times [0, x] \times [0, \infty] \times [0, \varepsilon_o])$ for some $\varepsilon_o, x_o > 0$, and if $f^{[0,0,-n,0]}(x, t, \infty, \varepsilon) = 0$ for all $n \geq 0$, then for any $N \geq 0$,*

$$y(x, t, \varepsilon) = \sum_{n=0}^{N} \varepsilon^n \phi^{[0,0,n]}(x, t/\varepsilon, 0) + O(\varepsilon^{N+1}) \tag{1.50}$$

uniformly for all $(x, t) \in [0, x_o] \times [0, x]$ as $\varepsilon \to 0^+$, where $\phi(x, T, \varepsilon) = f(x, \varepsilon T, T, \varepsilon)$.

1.7 Exercises

1.1. We noted in Section 1.2 that $y(x,\varepsilon) = f(x, x/\varepsilon)$, where $f(x,X) \in C^{\infty}([0,1] \times [0,\infty])$, implies $y(x,\varepsilon) = O_N y(x,\varepsilon) + O(\varepsilon^N)$ uniformly on $[\delta, 1]$ for any $\delta \in (0,1)$. In light of Theorem 1 this also means $[I_N - O_N I_N]y(x,\varepsilon) = O(\varepsilon^N)$ on $[\delta, 1]$. Show directly from (1.14) and (1.15) that in fact $[I_N - O_N I_N]y(x,\varepsilon) = O(\varepsilon^N)$ uniformly on $[\delta, c]$ for any $c > \delta > 0$.

1.2. Show that Corollary 1 applies to

$$y(x,\varepsilon) = \frac{\varepsilon + x^2}{\varepsilon + \varepsilon^2 + x + 2x^3} \qquad (1.51)$$

and that

$$O_3 I_3 y(x,\varepsilon) = x - (1 + 2x - 1/x)\varepsilon + (3 + 1/x - 1/x^2)\varepsilon^2. \qquad (1.52)$$

Note that this implies

$$u_0(x) = y_0(x), \quad u_1(x) = y_1(x) - 1/x, \quad u_2(x) = y_2(x) - 1/x + 1/x^2, \quad (1.53)$$

and

$$v_0(X) = Y_0(X), \quad v_1(X) = Y_1(X) + 1 - X, \quad v_2(X) = Y_2(X) - 3 + 2X. \quad (1.54)$$

Complete the calculation of $u_n(x)$ and $v_n(X)$ for $0 \le n \le 2$, and show that

$$\int_0^1 y(x,\varepsilon)\, dx = a + \varepsilon \ln(1/\varepsilon) - b\varepsilon + \varepsilon^2 \ln(1/\varepsilon) + c\varepsilon^2 - 4\varepsilon^3 \ln(1/\varepsilon) + O(\varepsilon^3), \quad (1.55)$$

where $a = \frac{1}{4}\ln 3$, $b = \frac{1}{6} + \frac{1}{2}\ln 3 + \frac{\sqrt{2}}{4}\arctan\sqrt{2}$, $c = -\frac{7}{18} - \frac{1}{2}\ln 3 + \frac{5\sqrt{2}}{4}\arctan\sqrt{2}$.

1.3. Complete the proof of Corollary 1 by considering

$$y_1(x,\varepsilon) = \sum_{n=0}^{N-1} \varepsilon^n u_n(x), \qquad y_2(x,\varepsilon) = \sum_{n=0}^{N-1} \varepsilon^n v_n(x/\varepsilon) \qquad (1.56)$$

separately. Clearly, for example, $I_N y_2(x,\varepsilon) = y_2(x,\varepsilon)$. Therefore $O_M y_2(x,\varepsilon) = O_M I_N y_2(x,\varepsilon)$. Also, $I_N O_M I_N y_2(x,\varepsilon) = O_M I_N y_2(x,\varepsilon)$, and so on.

1.4. Let m be a positive integer, assume $a, b > 0$ and suppose

$$F(\nu) = \nu^{m+1} \int_0^{\nu} t^m e^{-at^2} e^{-b(\nu t - t^2)}\, dt. \qquad (1.57)$$

Show that for $\nu > 0$,

$$F(\nu) = \varepsilon^{-1} \int_0^1 f(s, s/\varepsilon) \, ds, \tag{1.58}$$

where $\varepsilon = \nu^{-2}$, $f(s, S) = S^m e^{-Su(s)}$, $u(s) = as + b(1 - s)$. Since $u(s) > 0$ for $0 \le s \le 1$, it follows, as in Corollary 2, that for any $N \ge 0$,

$$f(s, s/\varepsilon) = \sum_{n=0}^{N-1} \varepsilon^n \phi^{[0,n]}(s/\varepsilon, 0) + O(\varepsilon^N) \tag{1.59}$$

as $\varepsilon \to 0^+$, where $\phi(S, \varepsilon) = f(\varepsilon S, S)$. Also,

$$\int_1^\infty \phi^{[0,n]}(s/\varepsilon, 0) \, ds = o(\varepsilon^\infty). \tag{1.60}$$

Show therefore that

$$F(\nu) = \sum_{n=0}^{N-1} c_n \nu^{-2n} + O(\nu^{-2N}) \tag{1.61}$$

as $\nu \to \infty$, where

$$c_n = (-1)^n \frac{(m + 2n)!}{n!} \frac{(a - b)^n}{b^{m+2n}}. \tag{1.62}$$

What if m is not an integer?

1.5. Consider the contour integral

$$F(\nu) = \int_C e^{-\nu h(z)} g(z) \, dz, \tag{1.63}$$

where C is the straight line from the origin to a point $z_o = r_o e^{i\theta_o}$ and $g(z)$, $h(z)$ are analytic on C. Assume $Re[h(z)] > 0$ on C, except $h^{[k]}(0) = 0$ for $0 \le k \le m - 1$ and θ_o is such that $\mu = Re[e^{im\theta_o} h^{[m]}(0)] > 0$, where $m \ge 1$. Observe that these assumptions imply the real part of $u(r) = r^{-m} h(r e^{i\theta_o})$ exceeds zero for $0 \le r \le r_o$, and therefore

$$F(\nu) = \int_0^{r_o} f(r, r/\varepsilon) \, dr, \tag{1.64}$$

where $\varepsilon = \nu^{-1/m}$ and $f(r, R) = g(r e^{i\theta_o}) \exp[-R^m u(r) + i\theta_o]$, which is in $C^\infty([0, r_o] \times [0, \infty))$, so we can apply Theorem 1. Furthermore, as in Exercise 1.4, $f^{[0,-n]}(r, \infty) = 0$ for all $n \ge 0$, and therefore

$$f(r, r/\varepsilon) = \sum_{n=0}^N \varepsilon^n \phi^{[0,n]}(r/\varepsilon, 0) + O(\varepsilon^{N+1}), \tag{1.65}$$

where $\phi(R, \varepsilon) = f(\varepsilon R, R)$. Also as in Exercise 1.4, $\phi^{[0,n]}(R, 0)$ is bounded by a polynomial times a decreasing exponential, in this case $\exp(-\mu R^m)$, and therefore

$$\int_{r_o}^{\infty} \phi^{[0,n]}(r/\varepsilon, 0)\, dr = o(\varepsilon^{\infty}). \tag{1.66}$$

Thus it follows from (1.64) that

$$F(\nu) = \varepsilon \sum_{n=0}^{N-1} \varepsilon^n \int_0^{\infty} \phi^{[0,n]}(R, 0)\, dR + O(\varepsilon^{N+1}). \tag{1.67}$$

In other words, if $\psi(Z, \varepsilon) = g(\varepsilon Z)e^{-h(\varepsilon Z)/\varepsilon^m}$, then

$$F(\nu) = \varepsilon \sum_{n=0}^{N-1} \varepsilon^n \int_0^{\infty e^{i\theta_o}} \psi^{[0,n]}(Z, 0)\, dZ + O(\varepsilon^{N+1}), \tag{1.68}$$

and $\psi^{[0,n]}(Z, 0)$ is a polynomial times $e^{-h^{[m]}(0)Z^m}$.

1.6. Beginning with the Bessel function integral

$$J_\nu(\nu) = \frac{1}{2\pi i} \int_{\infty - i\pi}^{\infty + i\pi} e^{-\nu h(z)}\, dz, \tag{1.69}$$

where $h(z) = z - \sinh z$, show that if $x_o > 0$, then

$$\int_{x_o \pm i\pi}^{\infty + i\pi} e^{-\nu h(z)}\, dz = o(\nu^{-\infty}) \tag{1.70}$$

as $\nu \to \infty$, and that therefore $J_\nu(\nu)$ is asymptotically equivalent to the sum of two integrals of the form assumed in the previous exercise, one with $\theta_o = \pi/3$, the other with $\theta_o = -\pi/3$, and both with $m = 3$. Show further that these two integrals can be combined to yield

$$J_\nu(\nu) = \varepsilon \sum_{n=0}^{N-1} \varepsilon^n \int_{\infty e^{-i\pi/3}}^{\infty e^{i\pi/3}} \psi^{[0,n]}(Z, 0)\, dZ + O(\varepsilon^{N+1}), \tag{1.71}$$

where $\varepsilon = \nu^{-1/3}$ and $\psi(Z, \varepsilon) = (1/2\pi i)\exp[-\varepsilon^{-3}h(\varepsilon Z)]$. Note that $\psi(Z, \varepsilon)$ is an even function of ε, so half the terms in (1.71) equal zero. If we let

$$a_n = \frac{1}{2\pi i} \int_{\infty e^{-i\pi/3}}^{\infty e^{i\pi/3}} Z^n e^{\frac{1}{6}Z^3}\, dZ, \tag{1.72}$$

then $a_0 = 2^{1/3}\text{Ai}(0)$, $a_1 = -2^{2/3}\text{Ai}^{[1]}(0)$, $a_2 = 0$, and $a_n = -2(n-2)a_{n-3}$ for $n \geq 3$. Use the Maple steps

$series(\varepsilon^{-3} \cdot (sinh(\varepsilon Z) - \varepsilon Z), \varepsilon = 0, 40):$
$series(e^{\%}, \varepsilon = 0, 40): convert(\%, polynom):$
$w := simplify(\% \cdot exp(-\frac{1}{6} \cdot Z^3)):$
$P := subs(\varepsilon = 1, \%): N := degree(P):$
$a_0 := A: a_1 := B: a_2 := 0: w := A - 1 + w:$
for n **from** 3 **to** N **do**
$a_n := -2 \cdot (n-2) \cdot a_{n-3}: w := subs(Z^n = a_n, w):$
end do: $w := series(w, \varepsilon = 0, 40);$

to determine the first 13 non-zero terms of (1.71).

1.7. Assume $a(x, \varepsilon) \in C^\infty([0,1] \times [0, \varepsilon_o])$. Show there exists $b(x, \varepsilon) \in C^\infty([0,1] \times [0, \varepsilon_o])$ and $c(\varepsilon) \in C^\infty([0, \varepsilon_o])$ such that, for $\varepsilon > 0$,

$$\frac{a(x, \varepsilon)}{x + \varepsilon} = b(x, \varepsilon) + \frac{c(\varepsilon)}{x + \varepsilon}. \tag{1.73}$$

Of course, this is trivial if $a(x, \varepsilon)$ is a polynomial.

1.8. An alternate proof of Theorem 1 can be obtained by a variation of the analysis in Section 2.1 of Part II of [3]. Note first that if $0 < \varepsilon^{1/2} \leq x \leq 1$, then $\varepsilon/x \leq \varepsilon^{1/2}$ and therefore, in view of (1.11), assuming $y(x, \varepsilon)$ satisfies the hypotheses of Theorem 1.1, $y(x, \varepsilon) = O_{2N}y(x, \varepsilon) + O(\varepsilon^N)$. Similarly, $I_N y(x, \varepsilon) = O_{2N}I_N y(x, \varepsilon) + O(\varepsilon^N)$ for $0 < \varepsilon^{1/2} \leq x \leq 1$. Therefore $y(x, \varepsilon) = C_N y(x, \varepsilon) + A_N y(x, \varepsilon) + O(\varepsilon^N)$ for $0 < \varepsilon^{1/2} \leq x \leq 1$, where $A_N y(x, \varepsilon) = [(O_{2N} - O_N) - I_N(O_{2N} - O_N)]y(x, \varepsilon)$. But a short calculation shows

$$A_N y(x, \varepsilon) = x^N \sum_{n=N}^{2N-1} (\varepsilon/x)^n \theta_n(x), \tag{1.74}$$

where

$$\theta_n(x) = x^{-N}\left[f^{[0, -n]}(x, \infty) - \sum_{m=0}^{N-1} x^m f^{[m, -n]}(0, \infty)\right], \tag{1.75}$$

and clearly $\theta_n(x) = O(1)$ for $0 \leq x \leq 1$. Therefore $A_N y(x, \varepsilon) = O(\varepsilon^N)$ for $0 < \varepsilon^{1/2} \leq x \leq 1$, since, in this case, $x^N(\varepsilon/x)^n \leq \varepsilon^N$ for $n \geq N$. An analogous argument, beginning with $y(x, \varepsilon) = I_{2N}y(x, \varepsilon) + O(\varepsilon^N)$, shows $y(x, \varepsilon) = C_N y(x, \varepsilon) + O(\varepsilon^N)$ as $\varepsilon \to 0^+$ when $0 \leq x \leq \varepsilon^{1/2}$.

Chapter 2
First Order Differential Equations

2.1 Problem A

The plan for this book is to apply the theory developed in Chapter 1 to a sequence of more or less increasingly complex differential equation problems. We begin with

$$\varepsilon y' + a(x, \varepsilon)y = b(x, \varepsilon), \tag{2.1}$$

where $a(x, \varepsilon), b(x, \varepsilon) \in C^\infty([0,1] \times [0, \varepsilon_o])$ for some $\varepsilon_o > 0$, and $a(x, \varepsilon) > 0$. We also assume $y(0, \varepsilon) = \alpha(\varepsilon) \in C^\infty([0, \varepsilon_o])$. The problem is to asymptotically approximate $y(x, \varepsilon)$ uniformly on $0 \leq x \leq 1$ as $\varepsilon \to 0^+$. If we put $z(x, \varepsilon) = y(x, \varepsilon) - \alpha(\varepsilon)$, then

$$z(x, \varepsilon) = \varepsilon^{-1} \int_0^x \hat{b}(x - t, \varepsilon)e^{-[k(x,\varepsilon) - k(x-t,\varepsilon)]/\varepsilon} \, dt, \tag{2.2}$$

where $\hat{b}(x, \varepsilon) = b(x, \varepsilon) - \alpha(\varepsilon)a(x, \varepsilon)$ and

$$k(x, \varepsilon) = \int_0^x a(t, \varepsilon) \, dt. \tag{2.3}$$

If we let $u(x, t, \varepsilon) = t^{-1}[k(x, \varepsilon) - k(x - t, \varepsilon)]$, then $u(x, t, \varepsilon) > 0$ on $[0, 1] \times [0, x] \times [0, \varepsilon_o]$. In particular, $u(x, 0, \varepsilon) = a(x, \varepsilon) > 0$. Thus we can rewrite (2.2) as

$$z(x, \varepsilon) = \varepsilon^{-1} \int_0^x f(x, t, t/\varepsilon, \varepsilon) \, dt, \tag{2.4}$$

where

$$f(x, t, T, \varepsilon) = \hat{b}(x - t, \varepsilon)e^{-Tu(x,t,\varepsilon)}, \tag{2.5}$$

and we can apply Corollary 2. In terms of $\phi(x, T, \varepsilon) = f(x, \varepsilon T, T, \varepsilon)$, it follows that

$$z(x, \varepsilon) = \sum_{n=0}^N \varepsilon^n \Phi_n(x, x/\varepsilon) + O(\varepsilon^N), \tag{2.6}$$

where

$$\Phi_n(x, X) = \int_0^X \phi^{[0,0,n]}(x, T, 0)\, dT.$$ (2.7)

Furthermore,

$$\phi^{[0,0,n]}(x, T, 0) = p_n(x, T)e^{-a(x,0)T},$$ (2.8)

where $p_n(x, T)$ is a polynomial in T, and therefore $\Phi_n(x, X) \in C^\infty([0,1] \times [0, \infty])$. Hence, in accordance with Corollary 1, applied to the sum in (2.6), we see $y(x, \varepsilon) = z(x, \varepsilon) + \alpha(\varepsilon)$ has a uniformly valid asymptotic expansion of the form

$$y(x, \varepsilon) = \sum_{n=0}^{N-1} \varepsilon^n [u_n(x) + v_n(x/\varepsilon)] + O(\varepsilon^N),$$ (2.9)

where $u_n(x) \in C^\infty([0,1])$, $v_n(X) \in C^\infty([0, \infty])$, $v_n(\infty) = 0$, and these terms can be found by computing inner and outer expansions.

From (2.6), we have $O_1 z(x, \varepsilon) = \Phi_0(x, \infty)$ and since $p_0(x, T) = \hat{b}(x, 0)$, it follows that

$$O_1 z(x, \varepsilon) = \hat{b}(x, 0)/a(x, 0).$$ (2.10)

Similarly,

$$I_1 z(x, \varepsilon) = \Phi_0(0, x/\varepsilon) = [\hat{b}(0, 0)/a(0, 0)](1 - e^{-a(0,0)x/\varepsilon})$$ (2.11)

and, from either (2.10) or (2.11),

$$O_1 I_1 z(x, \varepsilon) = I_1 O_1 z(x, \varepsilon) = \hat{b}(0, 0)/a(0, 0).$$ (2.12)

Also $C_1 y(x, \varepsilon) = C_1 z(x, \varepsilon) + \alpha(0)$. Hence, the first terms of (2.9) are

$$u_0(x) = b(x, 0)/a(x, 0), \quad v_0(X) = [\alpha(0) - b(0, 0)/a(0, 0)]e^{-a(0,0)X}.$$ (2.13)

Of course, it is much easier to calculate the inner and outer expansions for (2.9) directly from the differential equation (2.1). For the N-term outer expansion,

$$O_N y(x, \varepsilon) = \sum_{n=0}^{N-1} \varepsilon^n y_n(x),$$ (2.14)

(2.1) implies

$$y'_{n-1}(x) + \sum_{k=0}^n a_{n-k}(x)y_k(x) = b_n(x),$$ (2.15)

where $a_n(x) = a^{[0,n]}(x, 0)$, $b_n(x) = b^{[0,n]}(x, 0)$. Similarly, for the N-term inner expansion,

$$I_N y(x, \varepsilon) = \sum_{n=0}^{N-1} \varepsilon^n Y_n(x/\varepsilon),$$ (2.16)

if we put $A(X, \varepsilon) = a(\varepsilon X, \varepsilon)$, $B(X, \varepsilon) = b(\varepsilon X, \varepsilon)$, in addition to $Y(X, \varepsilon) = y(\varepsilon X, \varepsilon)$, then (2.1) becomes

$$Y' + A(X, \varepsilon)Y = B(X, \varepsilon). \tag{2.17}$$

Therefore,

$$Y'_n(X) + \sum_{k=0}^{n} A_{n-k}(X)Y_k(X) = B_n(X), \tag{2.18}$$

where $A_n(X) = A^{[0,n]}(X, 0)$, $B_n(X) = B^{[0,n]}(X, 0)$. Also, $y(0, \varepsilon) = \alpha(\varepsilon)$ implies $Y_n(0) = \alpha^{[n]}(0)$. For the actual calculations we offer our first Maple program.

```
ProbA := proc (a, b, α, N)
ONy := sum(εⁿ · yₙ, n = 0..N − 1);
ONdy := sum(εⁿ · dyₙ, n = 0..N − 1);
ONeq := series(ε · ONdy + a · ONy − b, ε = 0, N);
for k from 0 to N − 1 do
temp := coeff(ONeq, ε, k); yk := solve(temp = 0, yₖ);
dyk := diff(yk, x); ONeq := subs(yₖ = yk, dyₖ = dyk, ONeq);
ONy := subs(yₖ = yk, ONy); print(uₖ = yk);
end do;
A := subs(x = ε · X, a); B := subs(x = ε · X, b);
INy := sum(εⁿ · Yₙ, n = 0..N − 1);
INdy := sum(εⁿ · dYₙ, n = 0..N − 1);
INde := series(INdy + A * INy − B, ε = 0, N);
Nα := series(α, ε = 0, N);
for k from 0 to N − 1 do
temp := coeff(INde, ε, k);
de := subs(Yₖ = z(X), dYₖ = diff(z(X), X), temp) = 0;
dsolve({de, z(0) = coeff(Nα, ε, k)});
Yk := rhs(%); dYk := diff(Yk, X);
INde := subs(Yₖ = Yk, dYₖ = dYk, INde); Yₖ := Yk;
end do;
INONy := series(subs(x = ε · X, ONy), ε = 0, N);
for k from 0 to N − 1 do
vk := Yₖ − coeff(INONy, ε, k);
print(vₖ = simplify(expand(vk)));
end do;
end proc:
```

This program solves for the N-term outer expansion of $y(x, \varepsilon)$ using (2.15) and then computes the N-term inner expansion using (2.18). At the end, the program computes $I_N O_N y(x, \varepsilon)$. Normally there would be the matter of separating $I_N O_N y(x, \varepsilon)$ into two parts to form the functions $u_n(x)$ and $v_n(X)$.

In this problem, however, it is clear from (2.15) that $O_N y(x, \varepsilon)$ is continuous at $x = 0$ and therefore $u_n(x) = y_n(x)$. This also means $v_n(x/\varepsilon)$ is the coefficient of ε^n in $[I_N - I_N O_N]y(x, \varepsilon)$. As an example,

$$a := 2 + x + \varepsilon; \quad b := (2 + x)^{-1}; \quad \alpha := \varepsilon; \quad ProbA(a, b, \alpha, 2);$$

yields

$$u_0(x) = \frac{1}{(2 + x)^2}, \qquad u_1(x) = \frac{-x}{(2 + x)^4}, \tag{2.19}$$

and

$$v_0(X) = -\frac{1}{4}e^{-2X}, \qquad v_1(X) = \frac{1}{8}(8 + 2X + X^2)e^{-2X}. \tag{2.20}$$

2.2 Problem B

For our second differential equation we take

$$\varepsilon^2 y' + xa(x, \varepsilon)y = \varepsilon b(x, \varepsilon). \tag{2.21}$$

Again we assume $a(x, \varepsilon), b(x, \varepsilon) \in C^\infty([0, 1] \times [0, \varepsilon_o])$ for some $\varepsilon_o > 0$. We also assume $a(x, \varepsilon) > 0$ on $[0, 1] \times [0, \varepsilon_o]$, $y(0, \varepsilon) = \varepsilon\alpha(\varepsilon)$, where $\alpha(\varepsilon) \in C^\infty([0, \varepsilon_o])$, and, without loss of generality, we assume $a(0, 0) = 1$. The essential difference here from Problem A is the factor of x multiplying $a(x, \varepsilon)$.

In place of (2.3), we now have

$$k(x, \varepsilon) = \int_0^x ta(t, \varepsilon)\, dt \tag{2.22}$$

and therefore, as $t \to 0^+$,

$$k(x - t, 0) = k(x, 0) - txa(x, 0) + \frac{1}{2}t^2[a(x, 0) + xa^{[1,0]}(x, 0)] + O(t^3). \tag{2.23}$$

If we put

$$u(x, t, \varepsilon) = t^{-2}[k(x, \varepsilon) - k(x - t, \varepsilon)] - (tx - t^2)a(x, 0), \tag{2.24}$$

then $u(0, 0, 0) = 1/2$ and therefore $u(x, t, \varepsilon) > 0$ on $[0, x_o] \times [0, x] \times [0, \varepsilon_o]$ for some $x_o > 0$, and possibly a smaller $\varepsilon_o > 0$. Thus if we let $z(x, \varepsilon) = y(x, \varepsilon) - \varepsilon\alpha(\varepsilon)$ and $\hat{b}(x, \varepsilon) = b(x, \varepsilon) - x\alpha(\varepsilon)a(x, \varepsilon)$, then

$$z(x,\varepsilon) = \varepsilon^{-1} \int_0^x f(x,t,t/\varepsilon,\varepsilon)e^{-t(x-t)a(x,0)/\varepsilon^2}\, dt, \qquad (2.25)$$

where,

$$f(x,t,T,\varepsilon) = \hat{b}(x-t,\varepsilon)e^{-T^2 u(x,t,\varepsilon)}, \qquad (2.26)$$

and we have $f(x,t,T,\varepsilon) \in C^\infty([0,x_o] \times [0,x] \times [0,\infty] \times [0,\varepsilon_o])$. Also

$$0 < e^{-t(x-t)a(x,0)/\varepsilon^2} \le 1 \qquad (2.27)$$

for all $(x,t,\varepsilon) \in [0,1] \times [0,x] \times (0,\varepsilon_o]$. Hence, applying Corollary 2 with $\phi(x,T,\varepsilon) = f(x,\varepsilon T,T,\varepsilon)$, from (2.25) we get

$$z(x,\varepsilon) = \sum_{n=0}^N \varepsilon^n \Phi_n(x,x/\varepsilon) + O(\varepsilon^N) \qquad (2.28)$$

uniformly as $\varepsilon \to 0^+$ for $0 \le x \le x_o$, where

$$\Phi_n(x,X) = \int_0^X \phi^{[0,0,n]}(x,T,0)e^{-T(X-T)a(x,0)}\, dT. \qquad (2.29)$$

Also,

$$\phi^{[0,0,n]}(x,T,0) = p_n(x,T)e^{-T^2 u(x,0,0)}, \qquad (2.30)$$

where $p_n(x,T)$ is a polynomial in T.

From Exercise 1.4, it is clear that

$$F(x,X) = \int_0^X T^m e^{-T^2 u(x,0,0)} e^{-T(X-T)a(x,0)}\, dT, \qquad (2.31)$$

for any integer $m \ge 0$, is in $C^\infty([0,x_o] \times [0,\infty])$. Therefore $\Phi_n(x,X) \in C^\infty([0,x_o] \times [0,\infty])$ and thus from (2.28), by Corollary 1, we know $y(x,\varepsilon)$ has a uniformly valid expansion, at least for $0 \le x \le x_o$, of the form

$$y(x,\varepsilon) = \sum_{n=0}^{N-1} \varepsilon^n [u_n(x) + v_n(x/\varepsilon)] + O(\varepsilon^N). \qquad (2.32)$$

But also, from Problem A, we know that $y(x,\varepsilon) = O_N y(x,\varepsilon) + O(\varepsilon^N)$ uniformly as $\varepsilon \to 0^+$ for $x_o \le x \le 1$, since $xa(x,\varepsilon) > 0$ for $x_o \le x \le 1$. Furthermore, by Exercise 1.1, we know $[I_N - O_N I_N]y(x,\varepsilon) = O(\varepsilon^N)$ for $x_o \le x \le 1$. Hence, in conclusion, $y(x,\varepsilon)$ has a uniformly valid expansion of the form (2.32) as $\varepsilon \to 0^+$ on the full interval $0 \le x \le 1$, where $u_n(x) \in C^\infty([0,1])$, $v_n(X) \in C^\infty([0,\infty])$, $v_n(\infty) = 0$, and these functions can be determined by computing the corresponding outer and inner expansions for $y(x,\varepsilon)$ directly from the differential equation (2.21).

The inner expansion calculations for Problem B, if we just ask Maple to solve the associated sequence of differential equations, quickly gets bogged down with iterated error function integrals, so we have to help. It turns out that at each stage the equation to solve has the form

$$Y' + XY = \rho(X) + \sigma(X)P(X) + \tau(X)e^{-\frac{1}{2}X^2}, \qquad (2.33)$$

where $\rho(X)$, $\sigma(X)$, $\tau(X)$ are polynomials and

$$P(X) = e^{-\frac{1}{2}X^2} \int_0^X e^{\frac{1}{2}T^2}\, dT. \qquad (2.34)$$

The solution to this equation has the same form as its right hand side. That is,

$$Y(X) = p(X) + q(X)P(X) + r(X)e^{-\frac{1}{2}X^2}, \qquad (2.35)$$

where $p(X)$, $q(X)$, $r(X)$ are polynomials. Indeed, if we substitute (2.35) into (2.33), we find

$$p' + q + Xp = \rho(X), \qquad q' = \sigma(X), \qquad r' = \tau(X). \qquad (2.36)$$

Therefore

$$q(X) = q_0 + \int_0^X \sigma(T)\, dT, \quad r(X) = r_0 + \int_0^X \tau(T)\, dT, \qquad (2.37)$$

where $q_0 = q(0)$, $r_0 = r(0)$ have yet to be determined, and if d is the degree of $s(X) = \rho(X) - q(X)$, then, to satisfy (2.36a), the degree of $p(X)$ must be $d - 1$. Hence,

$$p(X) = \sum_{n=0}^{d-1} p_n X^n, \qquad s(X) = \sum_{n=0}^{d} s_n X^n \qquad (2.38)$$

and (2.36a), together with $p_d = p_{d+1} = 0$, implies

$$p_{d-k-1} = s_{d-k} - (d - k + 1)p_{d-k+1} \qquad (2.39)$$

for $0 \le k \le d-1$. Also $p_1 = s_0$ so $q_0 = \rho(0) - p_1$ and finally, $Y(0) = p_0 + r_0$ determines r_0. This is all incorporated into our Maple program for this problem.

```
ProbB := proc (a, b, α, N)
ONy := sum(εⁿ · yₙ, n = 0..N − 1);
ONdy := sum(εⁿ · dyₙ, n = 0..N − 1);
ONeq := series(ε² · ONdy + x · a · ONy − ε · b, ε = 0, N);
for k from 0 to N − 1 do
temp := coeff(ONeq, ε, k); yk := solve(temp = 0, yk);
dyk := diff(yk, x); ONeq := subs(yk = yk, dyk = dyk, ONeq);
```

```
ONy := subs(y_k = yk, ONy); y_k := yk;
end do;
A := subs(x = ε · X, a); B := subs(x = ε · X, b);
INy := sum(ε^n · Y_n, n = 0..N − 1);
Nεα := series(ε · α, ε = 0, N);
rths := series(B · I − X · A · INy + X · INy, ε = 0, N + 1);
for k from 0 to N − 1 do
temp := coeff(rths, ε, k);
qcut := int(subs(X = T, coeff(temp, P)), T = 0..X);
s := coeff(temp, I) − qcut;
if s = 0 then d := 0; else d := degree(s); end if;
p_d := 0; p_{d+1} := 0;
for j from 0 to d − 1 do
p_{d−j−1} := coeff(s, X, d − j) − (d − j + 1) · p_{d−j+1};
end do;
q := qcut + subs(X = 0, coeff(temp, I)) − p_1;
r := coeff(Nεα, ε, k) − p_0 + int(subs(X = T, coeff(temp, E)), T = 0..X);
p := sum(p_n · X^n, n = 0..d − 1);
Yk := p · I + q · P + r · E;
rths := subs(Y_k = Yk, rths); Y_k := Yk;
end do;
series(subs(x = ε · X, ONy), ε = 0, N); INONy := convert(%, polynom);
vpart := sum(X^n · coeff(INONy, X, n), n = 0..N);
upart := subs(X = ε^{−1} · x, INONy − vpart);
for k from 0 to N − 1 do
uk := y_k − coeff(upart, ε, k);
print(u_k = simplify(uk)); u_k := uk;
end do;
for k from 0 to N − 1 do
vk := Y_k − I · coeff(vpart, ε, k);
print(v_k = simplify(coeff(vk, I) · I + coeff(vk, P) · P + coeff(vk, E) · E));
end do;
end proc:
```

In this program, the letters I, P, and E are used to denote 1, $P(X)$, and $\exp(-\frac{1}{2}X^2)$, respectively. Also, we have used the fact that $I_N ON y(x, \varepsilon)$ is at most $O(X^{N-1})$ as $X \to \infty$ to split it into the two parts necessary to form $u_n(x)$ and $v_n(X)$ for $n = 0$ to $N − 1$.

As an example, if we set

$$a(x, \varepsilon) = 1 + cx^2, \qquad b(x, \varepsilon) = 1, \qquad \alpha(\varepsilon) = 0, \tag{2.40}$$

then $ProbB(a, b, \alpha, 4)$ yields

$$y(x, \varepsilon) = P(x/\varepsilon) - \frac{\varepsilon cx}{1 + cx^2} + \varepsilon^2 v_2(x/\varepsilon) + \varepsilon^3 u_3(x) + O(\varepsilon^4) \tag{2.41}$$

with

$$v_2(X) = \frac{c}{4}[X + X^3 + (3 - X^4)P(X)], \qquad u_3(x) = -\frac{c^2x(3 + cx^2)}{(1 + cx^2)^3}. \quad (2.42)$$

As another, if

$$a(x, \varepsilon) = 1 + gx, \qquad b(x, \varepsilon) = h + kx, \qquad \alpha(\varepsilon) = 0, \qquad (2.43)$$

then

$$u_0(x) = 0, \qquad u_1(x) = \frac{k - gh}{1 + gx}, \qquad v_0(X) = hP(X), \qquad (2.44)$$

and

$$v_1(X) = \frac{1}{3}gh(1 + X^2) - \frac{1}{3}ghX^3P(X) + \frac{1}{3}(2gh - 3k)e^{-\frac{1}{2}X^2}. \quad (2.45)$$

For

$$a(x, \varepsilon) = \cos(x) + x, \qquad b(x, \varepsilon) = \cos(x), \qquad \alpha(\varepsilon) = 1, \qquad (2.46)$$

a graph of

$$C_2y(x, \varepsilon) = P(x/\varepsilon) + \varepsilon[u_1(x) + v_1(x/\varepsilon)] \qquad (2.47)$$

when $\varepsilon = 0.2$ is shown in Figure 2.1, along with a portion of $O_2y(x, \varepsilon)$ and Maple's numerical solution of the differential equation. In this last example,

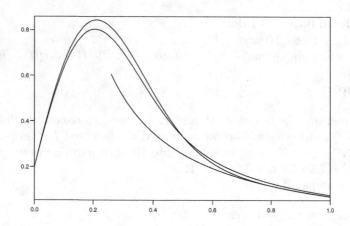

Fig. 2.1 Numerical solution of (2.21) and asymptotic approximations of the solution when $a(x, \varepsilon) = \cos(x) + x$, $b(x, \varepsilon) = \cos(x)$, $y(0, \varepsilon) = 1$ and $\varepsilon = 0.2$.

$$u_1(x) = \frac{-1}{\cos(x) + x}, \quad v_1(X) = \frac{1}{3}(1 + X^2) - \frac{1}{3}X^3 P(X) + \frac{5}{3}e^{-\frac{1}{2}X^2}. \quad (2.48)$$

2.3 Exercises

2.1. Let $y(x, \varepsilon)$ be the solution to

$$\varepsilon y' + x(1 + x + \varepsilon)y = 2 + x^2, \quad 1 \le x \le 2, \quad (2.49)$$

satisfying $y(1, \varepsilon) = 1 + \varepsilon$. Use *ProbA* to show

$$y(x, \varepsilon) = \frac{2 + x^2}{x(1 + x)} + e^{-2(x-1)/\varepsilon} + \varepsilon[u_1(x) + v_1((x-1)/\varepsilon)] + O(\varepsilon^2) \quad (2.50)$$

for $1 \le x \le 2$, where

$$u_1(x) = \frac{-2 - 4x + 3x^2 + 2x^3 + x^4 + x^5}{x^3(1 + x)^3}, \quad (2.51)$$

$$v_1(X) = -\frac{1}{2}(1 + 2X + 3X^2)e^{-2X}. \quad (2.52)$$

2.2. Note that $A_n(X)$ and $B_n(X)$ in (2.18) are polynomials and that therefore $Y_n(X) = p_n(X) + q_n(X)\exp[-a(0,0)X]$, where $p_n(X)$, $q_n(X)$ are polynomials. Therefore $v_n(X) = q_n(X)\exp[-a(0,0)X]$, since $v_n(\infty) = 0$. Furthermore,

$$v'_n(X) + \sum_{k=0}^{n} A_{n-k}(X)v_k(X) = 0, \quad (2.53)$$

and $v_n(0) = \alpha^{[n]}(0) - u_n(0)$. Write a new, shorter Maple program for computing the terms of (2.9).

2.3. Show that we could add ε^2 times $c(x, \varepsilon) \in C^\infty([0, 1] \times [0, \varepsilon_o])$, to $xa(x, \varepsilon)$ in (2.21) without upsetting the basic analysis, and modify the program *ProbB* appropriately to include it.

2.4. Let $y(x, \varepsilon)$ be the solution to

$$\varepsilon^2 y' + x(1 + x + \varepsilon)y = \varepsilon x(2 + x^2), \quad 0 \le x \le 1, \quad (2.54)$$

satisfying $y(0, \varepsilon) = 1$. Use *ProbB* to show

$$y(x, \varepsilon) = \frac{2 + x^2}{1 + x} - e^{-\frac{1}{2}(x/\varepsilon)^2} + \varepsilon[u_1(x) + v_1(x/\varepsilon)] + O(\varepsilon^2), \quad (2.55)$$

where

$$u_1(x) = -\frac{2+x^2}{(1+x)^2}, \quad v_1(X) = 2P(X) + \frac{1}{6}(12 + 3X^2 + 2X^3)e^{-\frac{1}{2}X^2}. \quad (2.56)$$

2.5. Suppose $s(x) \in C^\infty([0,1])$, $s'(x) > 0$ and $g(x,\varepsilon) \in C^\infty([0,1] \times [0,\varepsilon_o])$ for some $\varepsilon_o > 0$. Let

$$F(x,\varepsilon) = \varepsilon^{-1} \int_0^x e^{-[s(x)-s(t)]/\varepsilon} g(x,t)\, dt, \quad (2.57)$$

$$G(x,\varepsilon) = \varepsilon^{-1} \int_x^1 e^{[s(x)-s(t)]/\varepsilon} g(x,\varepsilon)\, dt. \quad (2.58)$$

From the analysis at the beginning of Section 2.1, we know $F(x,\varepsilon)$ has a uniformly valid expansion of the form

$$F(x,\varepsilon) = \sum_{n=0}^{N-1} \varepsilon^n [u_n(x) + v_n(x/\varepsilon)] + O(\varepsilon^N) \quad (2.59)$$

as $\varepsilon \to 0^+$, where $u_n(x) \in C^\infty([0,1])$, $v_n(X) \in C^\infty([0,\infty])$ and, as in Exercise 2.2, $v_n(X) = o(X^{-\infty})$ as $X \to \infty$. Show that $G(x,\varepsilon)$ has a uniformly valid expansion of the form

$$G(x,\varepsilon) = \sum_{n=0}^{N-1} \varepsilon^n [u_n(x) + w_n((1-x)/\varepsilon)] \quad (2.60)$$

as $\varepsilon \to 0^+$, where $u_n(x) \in C^\infty([0,1])$, $w_n(X) \in C^\infty([0,\infty])$ and $w_n(X) = o(X^{-\infty})$ as $X \to \infty$. In particular,

$$u_0(x) = g(x,0)/s'(x), \quad w_0(X) = e^{-s'(1)X}. \quad (2.61)$$

2.6. Suppose $g(x,t)$, $h(x,t) \in C^\infty([0,1] \times [0,x])$, $h(x,t) > 0$ for $0 < t \le x \le 1$, $h(x,0) = 0$ for $0 \le x \le 1$, and $h^{[0,1]}(x,0) > 0$ for $0 < x \le 1$ but $h^{[0,1]}(0,0) = 0$. Let

$$F(x,\nu) = \int_0^x e^{-\nu h(x,t)} g(x,t)\, dt. \quad (2.62)$$

We know that for $x > 0$,

$$F(x,\nu) = \nu^{-1}[g(x,0)/h^{[0,1]}(x,0)] + O(\nu^{-1}). \quad (2.63)$$

For an expansion of $F(x,\nu)$ that is uniformly valid for $0 \leq x \leq 1$, note first that as $(x,t) \to (0,0)$,

$$h(x,t) = h_{11}xt + h_{02}t^2 + O((x^2 + t^2)^{3/2}), \tag{2.64}$$

where we have begun using $h_{ij} = h^{[i,j]}(0,0)$, $g_{ij} = g^{[i,j]}(0,0)$, so it must be that $h_{11}x + h_{02}t \geq 0$ for $0 < t \leq x \leq 1$.

Assume $h_{11} > 0$ and $h_{02} > 0$. If we let

$$u(x,t) = t^{-2}[h(x,t) - th^{[0,1]}(x,0)], \qquad a(x) = x^{-1}h^{[0,1]}(x,0), \tag{2.65}$$

then $u(0,0) > 0$ and $a(0) > 0$, and hence there exists $x_o > 0$ such that $u(x,t) > 0$ on $[0,x_o] \times [0,x]$ and $a(x) > 0$ on $[0,x_o]$. Therefore

$$F(x,\nu) = \int_0^x f(x,t,t/\varepsilon)e^{-\nu x t a(x)}\, dt, \tag{2.66}$$

where $\varepsilon = \nu^{-1/2}$,

$$f(x,t,T) = e^{-T^2 u(x,t)}g(x,t) \tag{2.67}$$

and $f(x,t,T) \in C^\infty([0,x_o] \times [0,x] \times [0,\infty))$. In addition, $0 < e^{-\nu x t a(x)} \leq 1$ and thus, applying Corollary 2, with $\phi(x,T,\varepsilon) = f(x,\varepsilon T,T)$, we get

$$F(x,\nu) = \sum_{n=1}^{N-1} \varepsilon^n \Phi_n(x,x/\varepsilon) + O(\varepsilon^N) \tag{2.68}$$

uniformly as $\varepsilon \to 0^+$ for $0 \leq x \leq x_o$, where

$$\Phi_n(x,X) = \int_0^X \phi^{[0,0,n]}(x,T,0)e^{-XTa(x)}\, dT. \tag{2.69}$$

Show that $\Phi_n(x,X) \in C^\infty([0,x_o] \times [0,\infty))$, so we can apply Theorem 1 to each term of (2.68), and thereby determine

$$F(x,\nu) = \varepsilon v_1(x/\varepsilon) + \varepsilon^2[u_2(x) + v_2(x/\varepsilon)] + O(\varepsilon^3) \tag{2.70}$$

uniformly as $\varepsilon \to 0^+$ for $0 \leq x \leq x_o$, where

$$v_1(X) = g_{00} \int_0^X e^{-h_{02}T^2 - h_{11}XT}\, dT, \tag{2.71}$$

$$u_2(x) = \frac{g(x,0)}{h^{[0,1]}(x,0)} - \frac{g_{00}}{h_{11}x}, \tag{2.72}$$

and

$$v_2(X) = \int_0^X p_2(X,T)e^{-h_{02}T^2 - h_{11}XT}\, dT, \tag{2.73}$$

where

$$p_2(X,T) = g_{10} + (g_{01} - g_{00}h_{21}X^2)T - g_{00}h_{12}XT^2 - g_{00}h_{03}T^3. \qquad (2.74)$$

Note, furthermore, that for $x > 0$ the right side of (2.70) is asymptotically equal to $\varepsilon^2[g(x,0)/h^{[0,1]}(x,0)] + O(\varepsilon^3)$, the beginning of its outer expansion, and therefore, in view of (2.63), equation (2.70) actually holds uniformly for $0 \leq x \leq 1$.

The result (2.32) for the solution to Problem B is an example of an expansion for a case of $F(x,\nu)$ with $h_{11} > 0$ and $h_{02} < 0$.

Chapter 3
Second Order Differential Equations

3.1 Problem C

We begin this chapter with the classic singular perturbation problem

$$\varepsilon y'' + a(x,\varepsilon)y' + b(x,\varepsilon)y = c(x,\varepsilon), \tag{3.1}$$

where $a(x,\varepsilon) > 0$, subject to the boundary conditions $y(0,\varepsilon) = \alpha(\varepsilon)$ and $y(1,\varepsilon) = \beta(\varepsilon)$. It will be assumed that $a(x,\varepsilon)$, $b(x,\varepsilon)$, $c(x,\varepsilon) \in C^\infty([0,1] \times [0,\varepsilon_o])$ and $\alpha(\varepsilon)$, $\beta(\varepsilon) \in C^\infty([0,\varepsilon_o])$ for some $\varepsilon_o > 0$.

At first we will ignore the condition $y(1,\varepsilon) = \beta(\varepsilon)$ and instead treat (3.1) as an initial value problem with $y^{[1,0]}(0,\varepsilon) = \gamma(\varepsilon)/\varepsilon$ presumed given, where $\gamma(\varepsilon) \in C^\infty([0,\varepsilon_o])$, in addition to $y(0,\varepsilon) = \alpha(\varepsilon)$. From our Problem A analysis it is clear immediately that in the special case $b(x,\varepsilon) = a^{[1,0]}(x,\varepsilon)$ the solution to this initial value problem has the asymptotic form

$$y(x,\varepsilon) = \sum_{n=0}^{N-1} \varepsilon^n [u_n(x) + v_n(x/\varepsilon)] + \varepsilon^N R_N(x,\varepsilon), \tag{3.2}$$

where $u_n(x) \in C^\infty([0,1])$, $v_n(X) \in C^\infty([0,\infty])$, $v_n(X) = o(X^{-\infty})$ as $X \to \infty$ and $R_N(x,\varepsilon) = O(1)$ uniformly as $\varepsilon \to 0^+$ for $0 \le x \le 1$. We are going to prove that in fact this is true in general. First we will prove it for $N = 0$.

If we let $g(x,\varepsilon) = a^{[1,0]}(x,\varepsilon) - b(x,\varepsilon)$, then

$$\varepsilon y' + a(x,\varepsilon)y = f(x,\varepsilon) + \int_0^x g(t,\varepsilon)y(t,\varepsilon)\, dt, \tag{3.3}$$

where

$$f(x,\varepsilon) = \gamma(\varepsilon) + a(0,\varepsilon)\alpha(\varepsilon) + \int_0^x c(t,\varepsilon)\, dt. \tag{3.4}$$

Therefore, introducing

$$k(x, \varepsilon) = \int_0^x a(t, \varepsilon)\, dt, \tag{3.5}$$

as in Problem A, we have, after a reversal of integration order, the equivalent integral equation

$$y(x, \varepsilon) = \phi(x, \varepsilon) + \int_0^x K(x, t, \varepsilon) y(t, \varepsilon)\, dt, \tag{3.6}$$

where

$$\phi(x, \varepsilon) = \alpha(\varepsilon) + \varepsilon^{-1} \int_0^x e^{-[k(x,\varepsilon)-k(t,\varepsilon)]/\varepsilon} f(t, \varepsilon)\, dt \tag{3.7}$$

and

$$K(x, t, \varepsilon) = \varepsilon^{-1} g(t, \varepsilon) \int_t^x e^{-[k(x,\varepsilon)-k(s,\varepsilon)]/\varepsilon}\, ds. \tag{3.8}$$

Again as in Problem A, we know by Corollary 2 that

$$\varepsilon^{-1} \int_0^x e^{-[k(x,\varepsilon)-k(t,\varepsilon)]/\varepsilon}\, dt = O(1) \tag{3.9}$$

uniformly as $\varepsilon \to 0^+$ for $0 \le x \le 1$. Therefore, since $\alpha(\varepsilon) = O(1)$ and $f(x, \varepsilon)$ is uniformly $O(1)$ as $\varepsilon \to 0^+$ for $0 \le x \le 1$, there exists $M > 0$ such that $|\phi(x, \varepsilon)| \le M$ for all $(x, \varepsilon) \in [0,1] \times (0, \varepsilon_o]$. Similarly, $|K(x, t, \varepsilon)| \le M$ for all $(x, t, \varepsilon) \in [0,1] \times [0, x] \times (0, \varepsilon_o]$. Hence, by a simple Gronwall argument, $y(x, \varepsilon)$ is uniformly $O(1)$ as $\varepsilon \to 0^+$ for $0 \le x \le 1$. Indeed, for all $(x, \varepsilon) \in [0,1] \times (0, \varepsilon_o]$, we have

$$|y(x, \varepsilon)| \le \psi(x, \varepsilon) = M + M \int_0^x |y(t, \varepsilon)|\, dt, \tag{3.10}$$

so $\psi^{[1,0]}(x, \varepsilon) \le M\psi(x, \varepsilon)$ and therefore $e^{-Mx}\psi(x, \varepsilon) \le \psi(0, \varepsilon) = M$, or $|y(x, \varepsilon)| \le Me^{Mx} \le Me^M$.

It is easy to see what happens if we substitute (3.2) into (3.1). Let $a_n(x) = a^{[0,n]}(x, 0)$, $b_n(x) = b^{[0,n]}(x, 0)$ and $c_n(x) = c^{[0,n]}(x, 0)$. Also let $A(X, \varepsilon) = a(\varepsilon X, \varepsilon)$, $B(X, \varepsilon) = b(\varepsilon X, \varepsilon)$, $A_n(X) = A^{[0,n]}(X, 0)$ and $B_n(X) = B^{[0,n]}(X, 0)$. Note, for example, that

$$a(x, \varepsilon) \sum_{n=0}^{N-1} \varepsilon^n u_n'(x) = \sum_{n=0}^{N-1} \varepsilon^n \sum_{k=0}^n a_k(x) u_{n-k}'(x) + O(\varepsilon^N) \tag{3.11}$$

uniformly as $\epsilon \to 0^+$ for $0 \le x \le 1$. Similarly, since we are expecting $v_n(X) = o(X^{-\infty})$ as $X \to \infty$, Corollary 2 implies

$$a(x, \varepsilon) \sum_{n=0}^{N} \varepsilon^n v_n'(x/\varepsilon) = \sum_{n=0}^{N} \varepsilon^n \sum_{k=0}^n A_k(x/\varepsilon) v_{n-k}'(x/\varepsilon) + O(\varepsilon^{N+1}) \tag{3.12}$$

uniformly as $\varepsilon \to 0^+$ for $0 \leq x \leq 1$. Hence if we put

$$u''_{n-1}(x) + \sum_{k=0}^{n} a_k(x)u'_{n-k}(x) + \sum_{k=0}^{n} b_k(x)u_{n-k}(x) = c_n(x) \qquad (3.13)$$

for $0 \leq n \leq N - 1$ and

$$v''_n(X) + \sum_{k=0}^{n} A_k(X)v'_{n-k}(X) + \sum_{k=0}^{n-1} B_k(X)v_{n-1-k}(X) = 0, \qquad (3.14)$$

for $0 \leq n \leq N$, then (3.2) implies

$$\varepsilon r''_N + a(x,\varepsilon)r'_N + b(x,\varepsilon)r_N = \theta_N(x,\varepsilon), \qquad (3.15)$$

where $r_N(x,\varepsilon) = R_N(x,\varepsilon) - v_N(x/\varepsilon)$ and $\theta_N(x,\varepsilon)$ is uniformly $O(1)$ as $\varepsilon \to 0^+$ for $0 \leq x \leq 1$.

To satisfy $y(0,\varepsilon) = \alpha(\varepsilon)$ we require, according to (3.2), $u_n(0)+v_n(0) = \alpha_n$, where $\alpha_n = \alpha^{[n]}(0)$, and this implies $r_N(0,\varepsilon) = O(1)$. Also $y^{[1,0]}(0,\varepsilon) = \varepsilon^{-1}\gamma(\varepsilon)$ means $u'_{n-1}(0) + v'_n(0) = \gamma_n$, where $\gamma_n = \gamma^{[n]}(0)$, so $r_N^{[1,0]}(0,\varepsilon) = O(1)$. Therefore, just as $y(x,\varepsilon) = O(1)$, it follows that $r_N(x,\varepsilon) = O(1)$ uniformly as $\varepsilon \to 0^+$ for $0 \leq x \leq 1$. To complete the proof that the asymptotic solution of the initial value problem for (3.1) has the stipulated form (3.2), we need to confirm that $v_n(X) = o(X^{-\infty})$ as $X \to 0^+$. But this follows immediately from (3.14) once we assert $v_n(\infty) = 0$.

Note that to calculate the terms of (3.2) we have to alternate back and forth between (3.13) and (3.14), starting with (3.14), applying $v'_n(0) = \gamma_n - u'_{n-1}(0)$, $v_n(\infty) = 0$, first with $n = 0$, then going to (3.13) with $u_n(0) = \alpha_n - v_n(0)$, thereby determining $v'_1(0) = \gamma_1 - u'_0(0)$, and so on. For the boundary value problem, on the other hand, to satisfy $y(1,\varepsilon) = \beta(\varepsilon)$, note that (3.2) implies $u_n(1) = \beta^{[n]}(0)$, so $u_n(x)$ for $0 \leq n \leq N - 1$ can be determined separately. Then $v_n(X)$ for $0 \leq n \leq N - 1$ follows from (3.14) using $v_n(0) = \alpha_n - u_n(0)$, $v_n(\infty) = 0$. The fact that, indeed, the boundary value problem for (3.1) has a solution of the form (3.2) is confirmed by noting that the solution to the initial value problem has the form $y(x,\varepsilon) = w(x,\varepsilon) + \gamma(\varepsilon)z(x,\varepsilon)$, where $w(x,\varepsilon)$ is the solution to (3.1) satisfying $w(0,\varepsilon) = \alpha(\varepsilon)$, $w'(0,\varepsilon) = 0$ and $z(x,\varepsilon)$ is the solution to $\varepsilon z'' + a(x,\varepsilon)z' + b(x,\varepsilon)z = 0$ satisfying $z(0,\varepsilon) = 0$, $z'(0,\varepsilon) = 1/\varepsilon$. In particular, therefore, $z(1,0) = [1/a_0(0)] \exp \int_0^1 [-b_0(x)/a_0(x)]\, dx$, which is positive, so we can assume ε_o is such that $z(1,\varepsilon) > 0$ on $[0,\varepsilon_o]$, and thus the boundary value problem for (3.1) is equivalent to the initial value problem defined by taking $\gamma(\varepsilon) = [\beta(\varepsilon) - w(1,\varepsilon)]/z(1,\varepsilon)$. Note also that if, for the boundary value problem, we choose $\alpha(\varepsilon)$ such that $\alpha_n = u_n(0)$, then $v_n(X) = 0$. In other words, the N-term outer expansion,

$$O_N y(x, \varepsilon) = \sum_{n=0}^{N-1} \varepsilon^n u_n(x) \qquad (3.16)$$

is, by itself, the first N terms of a uniformly valid asymptotic expansion of a solution to (3.1).

Other strategies for calculating the terms of (3.2) are presented in the exercises. Below is a program for solving the boundary value problem based on the above strategy.

$ProbC := \mathbf{proc}(a, b, c, \alpha, \beta, N)$
$Nu := sum(\varepsilon^n \cdot u_n, n = 0..N - 1);\ Ndu := sum(\varepsilon^n \cdot du_n, n = 0..N - 1);$
$Nddu := sum(\varepsilon^n \cdot ddu_n, n = 0..N - 1);$
$Nude := series(\varepsilon \cdot Nddu + a \cdot Ndu + b \cdot Nu - c, \varepsilon = 0, N);$
$N\alpha := series(\alpha, \varepsilon, 0, N);\ N\beta := series(\beta, \varepsilon = 0, N);$
for k **from** 0 **to** $N - 1$ **do**
$temp := coeff(Nude, \varepsilon, k);$
$de := subs(u_k = z(x), du_k = diff(z(x), x), ddu_k = diff(z(x), x, x), temp);$
$bc := z(1) = coeff(N\beta, \varepsilon, k);$
$dsolve(\{de, bc\});\ uk := rhs(\%);\ duk := diff(uk, x);\ dduk := diff(duk, x);$
$Nude := subs(u_k = uk, du_k = duk, ddu_k = dduk, Nude);$
$print(u_k = simplify(uk));\ u_k := uk;$
end do;
$A := subs(x = \varepsilon \cdot X, a);\ B := subs(x = \varepsilon \cdot X, b);$
$Nv := sum(\varepsilon^n \cdot v_n, n = 0..N - 1);\ Ndv := sum(\varepsilon^n \cdot dv_n, n = 0..N - 1);$
$Nddv := sum(\varepsilon^n \cdot ddv_n, n = 0..N - 1);$
$Nvde := series(Nddv + A \cdot Ndv + \varepsilon \cdot B \cdot Nv, \varepsilon = 0, N);$
for k **from** 0 **to** $N - 1$ **do**
$temp := coeff(Nvde, \varepsilon, k);$
$de := subs(v_k = z(X), dv_k = diff(z(X), X), ddv_k = diff(z(X), X, X), temp);$
$bc := z(0) = coeff(N\alpha, \varepsilon, k) - subs(x = 0, u_k), z(\infty) = 0;$
$dsolve(\{de, bc\});\ vk := rhs(\%);\ dvk := diff(vk, X);\ ddvk := diff(dvk, X);$
$Nvde := subs(v_k = vk, dv_k = dvk, ddv_k = ddvk, Nvde);$
$print(v_k = simplify(vk));$
end do;
end proc:

3.2 Problem D

The problem for this section is

$$\varepsilon^2 y'' + \varepsilon^2 a(x, \varepsilon) y' - b(x, \varepsilon) y = c(x, \varepsilon) \qquad (3.17)$$

with $y(0, \varepsilon) = \alpha(\varepsilon)$, $y(1, \varepsilon) = \beta(\varepsilon)$. As usual, $a(x, \varepsilon)$, $b(x, \varepsilon)$, $c(x, \varepsilon) \in C^\infty(([0, 1] \times [0, \varepsilon_o])$ and $\alpha(\varepsilon)$, $\beta(\varepsilon) \in C^\infty([0, \varepsilon_o])$ for some $\varepsilon_o > 0$. In addition, in this problem we assume $b(x, \varepsilon) > 0$.

For the corresponding homogeneous differential equation, $c(x, \varepsilon) = 0$, if we put

$$y(x, \varepsilon) = e^{s(x)/\varepsilon} z(x, \varepsilon) \tag{3.18}$$

and choose $[s'(x)]^2 = b(x, 0)$, then

$$\varepsilon z'' + \hat{a}(x, \varepsilon) z' + \hat{b}(x, \varepsilon) z = 0, \tag{3.19}$$

where $\hat{a}(x, \varepsilon)$, $\hat{b}(x, \varepsilon) \in C^\infty([0, 1] \times [0, \varepsilon_o])$ and, in particular,

$$\hat{a}(x, \varepsilon) = 2s'(x) + \varepsilon a(x, \varepsilon). \tag{3.20}$$

If we take $s'(x)$ to be the positive square root of $b(x, 0)$, then (3.19) is a case of Problem C and, as noted at the end of Section 3.1, then (3.19) has a solution with a uniformly valid expansion as $\varepsilon \to 0^+$ of the form

$$z(x, \varepsilon) = \sum_{n=0}^{N-1} \varepsilon^n z_n(x) + O(\varepsilon^N), \tag{3.21}$$

where $z_n(x) \in C^\infty([0, 1])$. If we also require $z(0, 0) = 1$, then $z(0, \varepsilon) > 0$ for $\varepsilon > 0$ sufficiently small.

Actually, a solution to (3.19) of the form (3.21) also exists if $s'(x) = -[b(x, 0)]^{1/2}$. Indeed, in this case, simply replacing x by $1 - x$ changes (3.19) into another case of Problem C. Hence, to summarize, if we let

$$s(x) = \int_0^x [b(t, 0)]^{1/2} \, dt, \tag{3.22}$$

there exists a pair of linearly independent solutions

$$y_h^{(1)}(x, \varepsilon) = e^{-s(x)/\varepsilon} z^{(1)}(x, \varepsilon), \tag{3.23}$$

$$y_h^{(2)}(x, \varepsilon) = e^{-[s(1) - s(x)]/\varepsilon} z^{(2)}(x, \varepsilon), \tag{3.24}$$

to the homogeneous differential equation corresponding to (3.17), where for any $N > 0$, and certain $z_n^{(k)}(x) \in C^\infty([0, 1])$ with $z_0^{(k)}(0) = 1$,

$$z^{(k)}(x, \varepsilon) = \sum_{n=0}^{N-1} \varepsilon^n z_n^{(k)}(x) + O(\varepsilon^N) \tag{3.25}$$

uniformly as $\varepsilon \to 0^+$ for $0 \le x \le 1$.

The Wronskian of these two homogeneous equation solutions is $W(x, \varepsilon) = W(0, \varepsilon)\phi(x, \varepsilon)$, where

$$\phi(x,\varepsilon) = e^{-\int_0^x a(t,\varepsilon)\, dt} \tag{3.26}$$

and $W(0,\varepsilon) = \varepsilon^{-1} e^{-s(1)/\varepsilon} \Delta(\varepsilon)$, where $\Delta(\varepsilon)$ has an expansion in powers of ε as $\varepsilon \to 0^+$. In particular, $\Delta(0) = 2[b(0,0)]^{1/2}$. By the method of variation of parameters, if we let

$$g^{(1)}(x,t,\varepsilon) = z^{(1)}(x,\varepsilon)\psi(t,\varepsilon)z^{(2)}(t,\varepsilon), \tag{3.27}$$

$$g^{(2)}(x,t,\varepsilon) = z^{(2)}(x,\varepsilon)\psi(t,\varepsilon)z^{(1)}(t,\varepsilon), \tag{3.28}$$

where $\psi(t,\varepsilon) = c(t,\varepsilon)/[\Delta(\varepsilon)\phi(t,\varepsilon)]$, then the full differential equation (3.17) has a particular solution

$$y_p(x,\varepsilon) = F^{(1)}(x,\varepsilon) + F^{(2)}(x,\varepsilon), \tag{3.29}$$

where

$$F^{(1)}(x,\varepsilon) = \varepsilon^{-1} \int_0^x e^{-[s(x)-s(t)]/\varepsilon} g^{(1)}(x,t,\varepsilon)\, dt, \tag{3.30}$$

$$F^{(2)}(x,\varepsilon) = \varepsilon^{-1} \int_x^1 e^{-[s(t)-s(x)]/\varepsilon} g^{(2)}(x,t,\varepsilon)\, dt. \tag{3.31}$$

From Exercise 2.5, we know $F^{(1)}(x,\varepsilon)$, $F^{(2)}(x,\varepsilon)$ have uniformly valid asymptotic expansions for $0 \le x \le 1$, expressible as

$$F^{(1)}(x,\varepsilon) = \sum_{n=0}^{N-1} \varepsilon^n [u_n^{(1)}(x) + v_n^{(1)}(x/\varepsilon)] + O(\varepsilon^N), \tag{3.32}$$

$$F^{(2)}(x,\varepsilon) = \sum_{n=0}^{N-1} \varepsilon^n [u_n^{(2)}(x) + v_n^{(2)}((1-x)/\varepsilon)] + O(\varepsilon^N), \tag{3.33}$$

where $u_n^{(k)}(x) \in C^\infty([0,1])$, $v_n^{(k)}(X) \in C^\infty([0,\infty])$ and $v_n^{(k)}(X) = o(X^{-\infty})$ as $X \to \infty$. In addition, there exists $v_{h,n}^{(k)}(X) \in C^\infty([0,\infty])$ such that

$$y_h^{(1)}(x,\varepsilon) = \sum_{n=0}^{N-1} \varepsilon^n v_{h,n}^{(1)}(x/\varepsilon) + O(\varepsilon^N), \tag{3.34}$$

$$y_h^{(2)}(x,\varepsilon) = \sum_{n=0}^{N-1} \varepsilon^n v_{h,n}^{(2)}((1-x)/\varepsilon) + O(\varepsilon^N) \tag{3.35}$$

uniformly for $0 \le x \le 1$ as $\varepsilon \to 0^+$, and $v_{h,n}^{(k)}(X) = o(X^{-\infty})$ as $X \to \infty$.

Putting all this together we see the solution to (3.17), namely $y_p(x,\varepsilon)$ plus an ε-dependent linear combination of $y_h^{(1)}(x,\varepsilon)$ and $y_h^{(2)}(x,\varepsilon)$, subject to the boundary conditions $y(0,\varepsilon) = \alpha(\varepsilon)$, $y(1,\varepsilon) = \beta(\varepsilon)$, has a uniformly valid expansion for $0 \le x \le 1$ of the form

$$y(x, \varepsilon) = \sum_{n=0}^{N-1} \varepsilon^n [u_n(x) + v_n(x/\varepsilon) + w_n((1-x)/\varepsilon)] + O(\varepsilon^N) \qquad (3.36)$$

as $\varepsilon \to 0^+$, where $u_n(x) \in C^\infty([0,1])$ and $v_n(X)$, $w_n(X) \in C^\infty([0,\infty])$. Furthermore, as in Problem C, and also A, because we also have $v_n(X)$, $w_n(X) = o(X^{-\infty})$ as $X \to \infty$, each set of terms in this expansion can be determined separately. Indeed, with $a_n(x) = a^{[0,n]}(x, 0)$, $b_n(x) = b^{[0,n]}(x, 0)$ and $c_n(x) = c^{[0,n]}(x, 0)$, it is clear that

$$u''_{n-2}(x) + \sum_{k=0}^{n-2} a_k(x)u'_{n-2-k}(x) - \sum_{k=0}^{n} b_k(x)u_{n-k}(x) = c_n(x). \qquad (3.37)$$

Also, in terms of $A_n(X) = A^{[0,n]}(X, 0)$, $B_n(X) = B^{[0,n]}(X, 0)$, where $A(X, \varepsilon) = a(\varepsilon X, \varepsilon)$, $B(X, \varepsilon) = b(\varepsilon X, \varepsilon)$, we have

$$v''_n(X) + \sum_{k=0}^{n-1} A_k(X)v'_{n-1-k}(X) - \sum_{k=0}^{n} B_k(X)v_{n-k}(X) = 0, \qquad (3.38)$$

and finally, if we change and let $A(X, \varepsilon) = a(1 - \varepsilon X, \varepsilon)$, $B(X, \varepsilon) = b(1 - \varepsilon X, \varepsilon)$, then

$$w''_n(X) - \sum_{k=0}^{n-1} A_k(X)w'_{n-1-k}(X) - \sum_{k=0}^{n-1} B_k(X)w_{n-k}(X) = 0, \qquad (3.39)$$

where, again, $A_n(X) = A^{[0,n]}(X, 0)$, $B_n(X) = B^{[0,n]}(X, 0)$. So, first we find the outer expansion terms $u_n(x)$ for $0 \le n \le N - 1$, beginning with $u_0(x) = -c(x, 0)/b(x, 0)$, from (3.37). Then the (constant coefficient) differential equations (3.38) and (3.39) can be solved successively, with the boundary conditions $v_n(0) = \alpha^{[n]}(0) - u_n(0)$, $w_n(1) = \beta^{[n]}(0) - u_n(1)$, together with $v_n(\infty) = w_n(\infty) = 0$. A Maple program patterned after *ProbC* is readily devised to automate these calculations. To get more than a couple of terms, however, Maple needs assistance like the assistance provided in *ProbB* to calculate $v_n(X)$ and $w_n(X)$. This issue comes up in the next section, too.

If $a(x, \varepsilon) = 1 + x$, $b(x, \varepsilon) = 2 + x + \varepsilon$, $c(x, \varepsilon) = -4x^2$ and $\alpha(\varepsilon) = 2$, $\beta(\varepsilon) = 1/2$, then

$$u_0(x) = \frac{4x^2}{2+x}, \qquad v_0(X) = 2e^{-\sqrt{2}X}, \qquad w_0(X) = -\frac{5}{6}e^{-\sqrt{3}X}. \qquad (3.40)$$

Figure 3.1 is a graph of $u_0(x) + v_0(x/\varepsilon) + w_0((1-x)/\varepsilon)$ when $\varepsilon = .15$, together with Maple's numerical solution of (3.17) in this case and $u_0(x)$ alone. The same items are graphed in Figure 3.2 using $\varepsilon = .30$.

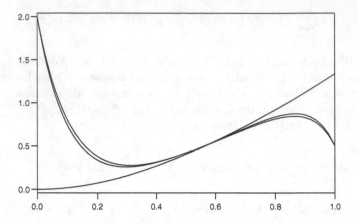

Fig. 3.1 Numerical solution to (3.17) and asymptotic approximations of the solution when $a(x, \varepsilon) = 1 + x$, $b(x, \varepsilon) = 2 + x + \varepsilon$, $c(x, \varepsilon) = -4x^2$, $y(0, \varepsilon) = 2$, $y(1, \varepsilon) = 1/2$ and $\varepsilon = 0.15$.

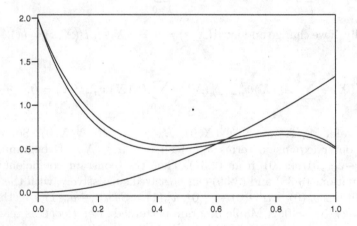

Fig. 3.2 Same as Figure 3.1 except $\varepsilon = 0.30$.

3.3 Problem E, Part 1

The problem for this section is a variation of Problem D in the same way Problem B is a variation of A. Assuming the functions $a(x, \varepsilon)$, $c(x, \varepsilon)$, $d(x, \varepsilon) \in C^{\infty}([0, 1] \times [0, \varepsilon_o])$ for some $\varepsilon_o > 0$, we also assume $0 < b(x) \in C^{\infty}([0, 1])$ and, without loss of generality, set $b(0) = 1$. The differential equation is

$$\varepsilon^2 y'' + \varepsilon^2 a(x, \varepsilon) y' - [xb(x) + \varepsilon^2 c(x, \varepsilon)] y = \varepsilon d(x, \varepsilon), \qquad (3.41)$$

and we are interested in the solution satisfying the boundary conditions $y(0, \varepsilon) = \alpha(\varepsilon)$, $y(1, \varepsilon) = \beta(\varepsilon)$, where $\alpha(\varepsilon)$, $\beta(\varepsilon) \in C^\infty([0, \varepsilon_o])$. We will start with an analysis of the special case

$$\varepsilon^2 y'' - xy = \varepsilon d(x, \varepsilon). \tag{3.42}$$

For (3.42), the corresponding homogeneous equation has the two linearly independent Airy function solutions

$$y_h^{(1)}(x, \varepsilon) = Ai(x/\varepsilon^{2/3}), \quad y_h^{(2)}(x, \varepsilon) = e^{-(2/3)/\varepsilon} Bi(x/\varepsilon^{2/3}) \tag{3.43}$$

and, knowing the Wronskian of $Ai(X)$ and $Bi(X)$ is $1/\pi$, it is readily verified that

$$y_p(x, \varepsilon) = F^{(1)}(x, \varepsilon) + F^{(2)}(x, \varepsilon) \tag{3.44}$$

is a particular solution to (3.42), where

$$F^{(1)}(x, \varepsilon) = -\pi Ai(x/\varepsilon^{2/3}) \int_0^x d(t, \varepsilon) Bi(t/\varepsilon^{2/3}) \, dt, \tag{3.45}$$

$$F^{(2)}(x, \varepsilon) = -\pi Bi(x/\varepsilon^{2/3}) \int_x^1 d(t, \varepsilon) Ai(t/\varepsilon^{2/3}) \, dt. \tag{3.46}$$

If we let

$$g(X) = e^{(2/3)X^{3/2}} Ai(X), \quad h(X) = e^{-(2/3)X^{3/2}} Bi(X), \tag{3.47}$$

and define $\hat{g}(X) = g(X^4)$, $\hat{h}(X) = h(X^4)$, then $\hat{g}(X)$, $\hat{h}(X) \in C^\infty([0, \infty])$. If we also let $\hat{x} = x^{1/4}$, $\hat{\varepsilon} = \varepsilon^{1/6}$, and substitute $t = \hat{t}^4$, then

$$F^{(1)}(x, \varepsilon) = \hat{g}(\hat{x}/\hat{\varepsilon}) \int_0^{\hat{x}} \hat{d}(\hat{t}, \hat{\varepsilon}) \hat{h}(\hat{t}/\hat{\varepsilon}) e^{-(2/3)(\hat{x}^6 - \hat{t}^6)/\hat{\varepsilon}^6} \, d\hat{t}, \tag{3.48}$$

$$F^{(2)}(x, \varepsilon) = \hat{h}(\hat{x}/\hat{\varepsilon}) \int_{\hat{x}}^1 \hat{d}(\hat{t}, \hat{\varepsilon}) \hat{g}(\hat{t}/\hat{\varepsilon}) e^{(2/3)(\hat{x}^6 - \hat{t}^6)/\hat{\varepsilon}^6} \, d\hat{t}, \tag{3.49}$$

where $\hat{d}(\hat{t}, \hat{\varepsilon}) = -4\pi \hat{t}^3 d(\hat{t}^4, \hat{\varepsilon}^6) \in C^\infty([0, 1] \times [0, \varepsilon_o^{1/6}])$. The asymptotic expansion form of these two integrals is readily established.

By Corollary 1, there exists $\phi_n(t) \in C^\infty([0, 1])$, $\psi_n(T) \in C^\infty([0, \infty])$ such that for any $N \geq 0$,

$$\hat{d}(\hat{t}, \hat{\varepsilon}) \hat{g}(\hat{t}/\hat{\varepsilon}) = \sum_{n=0}^{N-1} \hat{\varepsilon}^n [\phi_n(\hat{t}) + \psi_n(\hat{t}/\hat{\varepsilon})] + O(\hat{\varepsilon}^N) \tag{3.50}$$

uniformly for $0 \leq \hat{t} \leq 1$, and the same can be said for $\hat{d}(\hat{t}, \hat{\varepsilon}) \hat{h}(\hat{t}/\hat{\varepsilon})$. In turn, the results of Exercise 3.5, applied with $u(t) = \phi_n(t)$ and $v(T) = \psi_n(T)$, show there exists $\phi_n^{(k)}(x) \in C^\infty([0, 1])$, $\psi_n^{(k)}(X) \in C^\infty([0, \infty])$, and also $\theta_n^{(2)}(X) =$

$o(X^{-\infty})$ in $C^{\infty}([0,\infty])$, such that

$$F^{(1)}(x,\varepsilon) = \hat{g}(\hat{x}/\hat{\varepsilon}) \sum_{n=0}^{N-1} \hat{\varepsilon}^n [\phi_n^{(1)}(\hat{x}) + \psi_n^{(1)}(\hat{x}/\hat{\varepsilon})] + O(\hat{\varepsilon}^N), \qquad (3.51)$$

$$F^{(2)}(x,\varepsilon) = \hat{h}(\hat{x}/\hat{\varepsilon}) \sum_{n=0}^{N-1} \hat{\varepsilon}^n [\phi_n^{(2)}(\hat{x}) + \psi_n^{(2)}(\hat{x}/\hat{\varepsilon}) + \theta_n^{(2)}((1-x)/\varepsilon)] + O(\hat{\varepsilon}^N).$$

$$(3.52)$$

From here, with a few additional applications of Theorem 1, it is clear that the solution to (3.42) satisfying the boundary conditions $y(0,\varepsilon) = \alpha(\varepsilon)$, $y(1,\varepsilon) = \beta(\varepsilon)$ has the asymptotic form

$$y(x,\varepsilon) = \sum_{n=0}^{N-1} \hat{\varepsilon}^n [\hat{u}_n(\hat{x}) + \hat{v}_n(\hat{x}/\hat{\varepsilon}) + w_n((1-x)/\varepsilon)] + O(\hat{\varepsilon}^N), \qquad (3.53)$$

where $\hat{u}_n(x) \in C^{\infty}([0,1])$, both $\hat{v}_n(X), w_n(X) \in C^{\infty}([0,\infty])$, and $w_n(X) = o(X^{-\infty})$ as $X \to \infty$. For example, to deal with the contribution of $y_h^{(2)}(x,\varepsilon)$ to (3.53), let $t = 1 - x$ and observe that $y_h^{(2)}(x,\varepsilon) = f(t,t/\varepsilon,\hat{\varepsilon})$, where

$$f(t,T,\hat{\varepsilon}) = e^{-T\rho(t)}\hat{h}((1-t)^{1/4}/\hat{\varepsilon}) \qquad (3.54)$$

with $\rho(t) = (2/3)t^{-1}[1 - (1-t)^{3/2}]$. Since $\rho(t) > 0$ for $0 \le t \le 1/2$, $f(t,T,\hat{\varepsilon}) \in C^{\infty}([0,1/2] \times [0,\infty] \times [0,1])$. Thus we can apply Theorem 1 to each $f^{[0,0,n]}(t,t/\varepsilon,0)$ to see there exist polynomials $p_n(T)$ such that for any $N \ge 0$,

$$y_h^{(2)}(x,\varepsilon) = \sum_{n=0}^{N-1} \hat{\varepsilon}^n p_n(t/\varepsilon)e^{-t/\varepsilon} + O(\hat{\varepsilon}^N) \qquad (3.55)$$

uniformly as $\varepsilon \to 0^+$ for $0 \le t \le 1/2$. Furthermore, both $y^{(2)}(x,\varepsilon)$ and the sum in (3.55) are uniformly $o(\varepsilon^{\infty})$ for $1/2 \le t \le 1$. Hence (3.55) holds uniformly for the full interval, $0 \le x \le 1$.

As in Problem D, the homogeneous version of (3.41), the full differential equation for Problem E, has a pair of linearly independent solutions express-ible as

$$y_h^{(1)}(x,\varepsilon) = e^{-s(x)/\varepsilon}z^{(1)}(x,\varepsilon), \qquad (3.56)$$

$$y_h^{(2)}(x,\varepsilon) = e^{-[s(1)-s(x)]/\varepsilon}z^{(2)}(x,\varepsilon), \qquad (3.57)$$

where

$$s(x) = \int_0^x [tb(t)]^{1/2}\,dt. \qquad (3.58)$$

However, the asymptotic form of $z^{(k)}(x,\varepsilon)$ is more complex. It is shown in [9] that if $a(x,\varepsilon) = 0$, then for any $N \ge 0$,

$$z^{(k)}(x, \varepsilon) = \sum_{n=0}^{N-1} \varepsilon^{n/6}[u_n^{(k)}(x) + v_n^{(k)}(x/\varepsilon^{2/3})] + O(\varepsilon^{N/6}), \qquad (3.59)$$

where $u_n^{(k)}(x^4) \in C^\infty([0,1])$ and $v_n^{(k)}(X^4) \in C^\infty([0,\infty])$. In Exercise 3.6, we show that in fact this is true whether $a(x,\varepsilon) = 0$ or not. Therefore, as above for (3.42) and as in Problem D, we can use the method of variation of parameters to obtain a particular solution of (3.41). It is also a straightforward matter to determine the asymptotic form of this solution, and the homogeneous equation solutions (3.56) and (3.57).

Note that $s(x) = x^{3/2}\sigma(x)$, where $0 < \sigma(x) \in C^\infty([0,1])$, and again introduce $\hat{x} = x^{1/4}$ to get $s(x) = \hat{x}^6\hat{\sigma}(\hat{x})$, where $\hat{\sigma}(\hat{x}) = \sigma(\hat{x}^4)$. Then, for example,

$$e^{-s(x)/\varepsilon}v_n^{(k)}(x/\varepsilon^{2/3}) = f(\hat{x}, \hat{x}/\hat{\varepsilon}) \qquad (3.60)$$

where $\hat{\varepsilon} = \varepsilon^{1/6}$, as above, and

$$f(\hat{x}, \hat{X}) = e^{-\hat{X}^6\hat{\sigma}(\hat{x})}v_n^{(k)}(\hat{X}^4), \qquad (3.61)$$

which is in $C^\infty([0,1] \times [0,\infty])$. Furthermore, $f^{[0,-n]}(\hat{x}, \infty) = 0$, so

$$f(\hat{x}, \hat{x}/\hat{\varepsilon}) = \sum_{n=0}^{N-1} \hat{\varepsilon}^n \phi^{[0,n]}(\hat{x}/\hat{\varepsilon}, 0) + O(\hat{\varepsilon}^N), \qquad (3.62)$$

where $\phi(\hat{X}, \hat{\varepsilon}) = f(\hat{\varepsilon}\hat{X}, \hat{X})$. Also for example, to deal with

$$F(x, \varepsilon) = e^{-(\hat{x}/\hat{\varepsilon})^6\hat{\sigma}(\hat{x})} \int_0^{\hat{x}} u(\hat{t})e^{(\hat{t}/\hat{\varepsilon})^6\hat{\sigma}(\hat{t})}\, d\hat{t}, \qquad (3.63)$$

where $u(\hat{t}) \in C^\infty([0,1])$, let $\tilde{t} = \hat{t}[\hat{\sigma}(\hat{t})]^{1/6}$, let $\hat{t} = \theta(\tilde{t})$ denote the inverse of this transformation, and let $\tilde{x} = \hat{x}[\hat{\sigma}(\hat{x})]^{1/6}$ to obtain

$$F(x, \varepsilon) = e^{-(\tilde{x}/\hat{\varepsilon})^6} \int_0^{\tilde{x}} g(\tilde{t})e^{(\tilde{t}/\hat{\varepsilon})^6}\, d\tilde{t}, \qquad (3.64)$$

where $g(\tilde{t}) = u(\theta(\tilde{t}))\theta'(\tilde{t})$. It follows by Exercise 3.5 that

$$F(x, \varepsilon) = \sum_{n=0}^{N-1} \hat{\varepsilon}^n[\phi_n(\tilde{x}) + \psi_n(\tilde{x}/\hat{\varepsilon})] + O(\hat{\varepsilon}^N), \qquad (3.65)$$

for any $N \geq 0$, where $\phi_n(\tilde{x}) \in C^\infty([0,1])$ and $\psi_n(\tilde{X}) \in C^\infty([0,\infty])$. Also, we can apply Theorem 1 to $f(\hat{x}, \hat{X}) = \psi_n(\hat{X}[\hat{\sigma}(\hat{x})]^{1/6})$. Thus, ultimately, we see the solution to the boundary value problem for (3.41) has the same asymptotic form as the solution for the special case (3.42). That is, there exists $u_n(x)$, $v_n(X)$ and $w_n(X)$ such that, for any $N \geq 0$,

$$y(x,\varepsilon) = \sum_{n=0}^{N-1} \varepsilon^{n/6}[u_n(x) + v_n(x/\varepsilon^{2/3}) + w_n((1-x)/\varepsilon)] + O(\varepsilon^{N/6}), \quad (3.66)$$

uniformly as $\varepsilon \to 0^+$ for $0 \le x \le 1$, where $u_n(x^4) \in C^\infty([0,1])$, both $v_n(X^4)$ and $w_n(X)$ are in $C^\infty([0,\infty])$, $v_n(\infty) = 0$ and $w_n(X) = o(X^{-\infty})$ as $X \to \infty$.

3.4 Problem E, Part 2

To determine the functions $u_n(x) = u_n(\hat{x}^4) = \hat{u}_n(\hat{x})$ and $v_n(X) = v_n(\hat{X}^4) = \hat{v}_n(\hat{X})$ in (3.66), we need to compute the outer and inner expansions

$$O_N y(x,\varepsilon) = \sum_{n=0}^{N-1} \hat{\varepsilon}^n \hat{y}^{[0,n]}(\hat{x},0), \quad (3.67)$$

and

$$I_N y(x,\varepsilon) = \sum_{n=0}^{N-1} \hat{\varepsilon}^n \hat{Y}^{[0,n]}(\hat{X},0), \quad (3.68)$$

where $\hat{y}(\hat{x},\hat{\varepsilon}) = y(\hat{x}^4,\hat{\varepsilon}^6)$ and $\hat{Y}(\hat{X},\hat{\varepsilon}) = \hat{y}(\hat{\varepsilon}\hat{X},\hat{\varepsilon})$. If we let $z(x,\hat{\varepsilon}) = y(x,\hat{\varepsilon}^6)$, then $\hat{y}(\hat{x},\hat{\varepsilon}) = z(\hat{x}^4,\hat{\varepsilon})$ and therefore

$$O_N y(x,\varepsilon) = \sum_{n=0}^{N-1} \hat{\varepsilon}^n z_n(x) \quad (3.69)$$

where $z_n(x) = z^{[0,n]}(x,0)$. If we also let $Z(X,\hat{\varepsilon}) = y(\hat{\varepsilon}^4 X, \hat{\varepsilon}^6)$, then $Z(\hat{X}^4,\hat{\varepsilon}) = \hat{Y}(\hat{X},\hat{\varepsilon})$, so

$$I_N y(x,\varepsilon) = \sum_{n=0}^{N-1} \hat{\varepsilon}^n Z_n(X), \quad (3.70)$$

where $Z_n(X) = Z^{[0,n]}(X,0)$, and $X = \hat{X}^4$.

Upon substituting $\varepsilon = \hat{\varepsilon}^6$ in (3.41), it is clear that $z_n(x) = 0$ unless $n \ge 1$ is a multiple of 6. In particular,

$$z_6(x) = -d(x,0)/[xb(x)], \quad z_{12}(x) = -d^{[0,1]}(x,0)/[xb(x)]. \quad (3.71)$$

Also, therefore, from the expansion of $\hat{\varepsilon}^6 z_6(\hat{\varepsilon}^4 X) + \hat{\varepsilon}^{12} z_{12}(\hat{\varepsilon}^4 X)$, we have

$$I_{14} O_{14} y(x,\varepsilon) = -\hat{\varepsilon}^6[d_{00}/x + d_{10} - d_{00}b_1] - \hat{\varepsilon}^{10}[d_{00}(b_1^2 - b_2)$$

$$-d_{10}b_1 + d_{20}]X - \hat{\varepsilon}^{12}[d_{01}/x + d_{11} - d_{01}b_1], \quad (3.72)$$

where $d_{mn} = d^{[m,n]}(0,0)$ and $b_n = b^{[n]}(0)$. Hence,

$$u_6(x) = z_6(x) + d_{00}/x, \quad u_{12}(x) = z_{12}(x) + d_{01}/x \qquad (3.73)$$

and otherwise $u_n(x) = 0$ for $0 \le n \le 13$. Also,

$$v_6(X) = Z_6(X) + d_{10} - d_{00}b_1, \quad v_{12}(X) = Z_{12}(X) + d_{11} - d_{01}b_1, \qquad (3.74)$$

$$v_{10}(X) = Z_{10}(X) + [d_{00}(b_1^2 - b_2) - d_{10}b_1 + d_{20}]X, \qquad (3.75)$$

and otherwise $v_n(X) = Z_n(X)$ for $0 \le n \le 13$.

From the differential equation for $Z(X, \hat{\varepsilon})$ it is clear that $Z_n(X) = 0$ if n is odd. Also,

$$Z_0'' - XZ_0 = 0, \quad Z_0(0) = \alpha_0, \qquad (3.76)$$

$$Z_2'' - XZ_2 = d_{00}, \quad Z_2(0) = 0, \qquad (3.77)$$

$$Z_4'' - XZ_4 = b_1 X^2 Z_0 - a_{00} Z_0', \quad Z_4(0) = 0, \qquad (3.78)$$

$$Z_6'' - XZ_6 = b_1 X^2 Z_2 - a_{00} Z_2' - d_{10}X, \quad Z_6(0) = \alpha_1, \qquad (3.79)$$

where $\alpha_n = \alpha^{[n]}(0)$. In addition, $v_n(\infty) = 0$ implies, in particular, $Z_0(\infty) = Z_2(\infty) = 0$. Therefore

$$Z_0(X) = [\alpha_0/Ai(0)]Ai(X), \qquad (3.80)$$

and, in terms of

$$P(X) = -\pi Ai(X) \int_0^X Bi(T)\, dT - \pi Bi(X) \int_X^\infty Ai(T)\, dT, \qquad (3.81)$$

which satisfies $P'' - XP = 1$, we have

$$Z_2(X) = d_{00}P(X) - [d_{00}P(0)/Ai(0)]Ai(X). \qquad (3.82)$$

In general, to find $Z_n(X)$, we need to solve an equation of the form

$$Z'' - XZ = \lambda(X)R(X) + \mu(X)S(X) + \rho(X)P(X) + \sigma(X)Q(X) + \tau(X), \qquad (3.83)$$

where $R(X) = Ai(X)$, $S(X) = Ai'(X)$, $Q(X) = P'(X)$, and $\lambda(X)$, $\mu(X)$, $\rho(X)$, $\sigma(X)$ and $\tau(X)$ are polynomials. But the desired solution to an equation of this form has the same form as its right side. That is,

$$Z(X) = p(X)R(X) + q(X)S(X) + r(X)P(X) + s(X)Q(X) + t(X), \qquad (3.84)$$

where $p(X)$, $q(X)$, $r(X)$, $s(X)$ and $t(X)$ are polynomials.

Suppose, for example, that

$$\lambda(X) = \lambda_0 + \lambda_1 X + \lambda_2 X^2, \quad \mu(X) = \mu_0 + \mu_1 X + \mu_2 X^2. \qquad (3.85)$$

Substitution of (3.84) into (3.83) reveals

$$p'' - 2Xq' + q = \lambda(X), \quad q'' + 2p' = \mu(X). \tag{3.86}$$

and upon inserting

$$p(X) = p_0 + p_1 X + p_2 X^2 + p_3 X^3, \quad q(X) = q_0 + q_1 X + q_2 X^2, \tag{3.87}$$

we get

$$2p_2 + q_0 = \lambda_0, \quad 6p_3 + 3q_1 = \lambda_1, \quad 5q_2 = \lambda_2, \tag{3.88}$$

$$2q_2 + 2p_1 = \mu_0, \quad 4p_2 = \mu_0, \quad 6p_3 = \mu_2. \tag{3.89}$$

Hence, in order,

$$q_2 = \frac{1}{5}\lambda_2, \quad p_3 = \frac{1}{6}\mu_2, \quad q_1 = \frac{1}{3}(\lambda_1 - 6p_3), \tag{3.90}$$

$$p_2 = \frac{1}{4}\mu_0, \quad q_0 = \lambda_0 - 2p_2, \quad p_1 = \frac{1}{2}(\mu_0 - 2q_2). \tag{3.91}$$

The functions $r(X)$, $s(X)$, and $t(X)$ are similarly determined once $\rho(X)$, $\sigma(X)$ and $\tau(X)$ are given and, in the end, p_0 is determined by

$$Z(0) = p_0 R(0) + q_0 S(0) + r(0)P(0) + s(0)Q(0) + t(0). \tag{3.92}$$

Below is a Maple program that performs these calculations for us. The program also computes $w_n(X)$, using the fact that $w_n(X)$ is a polynomial times $\exp[-b(1)X]$, and

$$y(1,\varepsilon) = O_N y(1,\varepsilon) + \sum_{n=0}^{N-1} \hat{\varepsilon}^n w_n(0) + O(\hat{\varepsilon}^N), \tag{3.93}$$

which follows from (3.66), since, of course, $[I_N - O_N I_N]y(1,\varepsilon) = O(\hat{\varepsilon}^N)$. Although somewhat longer than our previous programs, *ProbE* follows a familiar procedure. First there is a section for the computation of $O_N y(x,\varepsilon)$. Then there is a longer than normal section, because of all the polynomials to determine, for the computation of $I_N y(x,\varepsilon)$. After computing $I_N O_N y(x,\varepsilon)$, the functions $u_n(x)$ and $v_n(X)$ are determined, and at the end of the program there is a separate section to determine each $w_n(X)$. The symbols I, P, Q, R and S in the program are used to denote 1, $P(X)$, $Q(X)$, $R(X)$ and $S(X)$, respectively. Also, $Po = P(0)$, $Qo = Q(0)$, $Ro = R(0)$ and $So = S(0)$.

```
ProbE := proc(a, b, c, d, α, β, N)
ONy := sum(ε^n · y_n, n = 0..N − 1);
ONdy := sum(ε^n · dy_n, n = 0..N − 1);
ONddy := sum(ε^n · ddy_n, n = 0..N − 1);
eq := subs(ε = ε^6, ε^2 · ONddy + ε^2 · a · ONdy − (x · b + ε^2 · c) · ONy − ε · d);
ONeq := series(eq, ε = 0, N);
for k from 0 to N − 1 do
```

$temp := coeff(ONeq, \epsilon, k); yk := solve(temp = 0, y_k);$
$dyk := diff(yk, x); ddyk := diff(dyk, x);$
$ONeq := subs(y_k = yk, dy_k = dyk, ddy_k = ddyk, ONeq);$
$ONy := subs(y_k = yk, ONy); y_k := yk;$
end do;
$N\alpha := series(subs(\varepsilon = \epsilon^6, \alpha), \epsilon = 0, N);$
$N\beta := series(subs(\varepsilon = \epsilon^6, \beta), \epsilon = 0, N);$
$A := subs(\varepsilon = \epsilon^6, x = \epsilon^4 \cdot X, a); B := subs(x = \epsilon^4 \cdot X, b);$
$C := subs(\varepsilon = \epsilon^6, x = \epsilon^4 \cdot X, c); D := subs(\varepsilon = \epsilon^6, x = \epsilon^4 \cdot X, d);$
$INy := sum(\epsilon^n \cdot Y_n, n = 0..N - 1); INdy := sum(\epsilon^n \cdot dY_n, n = 0..N - 1);$
$rths := series(-\epsilon^4 \cdot A \cdot INdy + (X \cdot B - X + \epsilon^8 \cdot C) \cdot INy + \epsilon^2 \cdot D \cdot I, \epsilon = 0, N+1);$
for k **from** 0 **to** $N - 1$ **do**
$temp := coeff(rths, \epsilon, k);$
if $temp = 0$ **then** $M := 0;$ **else** $M := degree(temp, X);$ **end if;**
$\lambda := coeff(temp, R); \mu := coeff(temp, S); \rho := coeff(temp, P);$
$\sigma := coeff(temp, Q); \tau := coeff(temp, I);$
$pm := sum(p_n \cdot X^n, n = 0..M + 1); qm := sum(q_n \cdot X^n, n = 0..M);$
$rm := sum(r_n \cdot X^n, n = 0..M + 1); sm := sum(s_n \cdot X^n, n = 0..M);$
$tm := sum(t_n \cdot X^n, n = 0..M);$
for j **from** 0 **to** M **do**
$coeff(diff(pm, X, X) + 2 \cdot X \cdot diff(qm, X) + qm - \lambda, X, M - j);$
$solve(\% = 0, q_{M-j}); qm := subs(q_{M-j} = \%, qm);$
$coeff(diff(qm, X, X) + 2 \cdot diff(pm, X) - \mu, X, M - j);$
$solve(\% = 0, p_{M+1-j}); pm := subs(p_{M+1-j} = \%, pm);$
$coeff(diff(rm, X, X) + 2 \cdot X \cdot diff(sm, X) + sm - \rho, X, M - j);$
$solve(\% = 0, s_{M-j}); sm := subs(s_{M-j} = \%, sm);$
$coeff(diff(sm, X, X) + 2 \cdot diff(rm, X) - \sigma, X, M - j);$
$solve(\% = 0); rm := subs(r_{M+1-j} = \%, rm);$
$coeff(diff(tm, X, X) - X \cdot tm + 2 \cdot diff(sm, X) + rm - \tau, X, M + 1 - j);$
$solve(\% = 0, t_{M-j}); tm := subs(t_{M-j} = \%, tm);$
end do;
$r0 := coeff(\tau, X, 0) - 2 \cdot coeff(sm, X, 1) - 2 \cdot coeff(tm, X, 2);$
$rm := subs(r_0 = r0, rm); r_0 := r0;$
$q0 := coeff(qm, X, 0); s0 := coeff(sm, X, 0); t0 := coeff(tm, X, 0);$
$p0 := (coeff(N\alpha, \epsilon, k) - q0 \cdot So - r0 \cdot Po - s0 \cdot Qo - t0) \cdot Ro^{-1};$
$p0 := simplify(\%); pm := subs(p_0 = p0, pm);$
$Yk := pm \cdot R + qm \cdot S + rm \cdot P \cdot sm \cdot Q + tm \cdot I;$
$dYk := (diff(pm, X) + X \cdot qm) \cdot R + (diff(qm, X) + pm) \cdot S$
$+ (diff(rm, X) + X \cdot sm) \cdot P + (diff(sm, X) + rm) \cdot Q + (diff(tm, X) + sm) \cdot I;$
$rths := subs(Y_k = Yk, dYk = dYk, rths); Y_k := Yk;$
end do;
$series(subs(x = \epsilon^4 \cdot X^4, ONy), \epsilon = 0, N); INONy := convert(\%, polynom);$
$vpart := sum(X^n \cdot coeff(INONy, X, n), n = 0..N);$
$upart := subs(X = \epsilon^{-4} \cdot x, INONy - vpart);$
for k **from** 0 **to** $N - 1$ **do**

$uk := y_k - coeff(upart, \epsilon, k);$
$print(u_k = simplify(uk));$
end do;
for k **from** 0 **to** $N - 1$ **do**
$vk := Y_k - I \cdot coeff(vpart, \epsilon, k);$
$print(v_k = vk);$
end do;
$A := subs(\varepsilon = \epsilon^6, x = 1 - \epsilon^6 \cdot X, a);$
$B := subs(\varepsilon = \epsilon^6, x = 1 - \epsilon^6 \cdot X, b);$
$C := subs(\varepsilon = \epsilon^6, x = 1 - \epsilon^6 \cdot X, c); \ m := sqrt(subs(x = 1, b));$
$Nw := sum(\epsilon^n \cdot w_n, n = 0..N - 1); \ Ndw := sum(\epsilon^n \cdot dw_n, n = 0..N - 1);$
$rths := series(-\epsilon^6 \cdot A \cdot Ndw + ((1 - \epsilon^6 \cdot X) \cdot B + \epsilon^{12} \cdot C - m^2) \cdot Nw, \epsilon = 0, N);$
for k **from** 0 **to** $N - 1$ **do**
$\chi := coeff(rths, \epsilon, k);$
if $\chi = 0$ **then** $K = 0$ **else** $K := degree(\chi, X);$ **end if;**
$\pi_0 := coeff(N\beta, \epsilon, k) - subs(x = 1, y_k); \ \pi_{K+2} := 0;$
for j **from** 0 **to** K **do**
$i := K - j;$
$\pi_{i+1} := ((i + 2) \cdot (i + 1) \cdot \pi_{i+2} - coeff(\chi, X, i)) \cdot (2 \cdot m \cdot (i + 1))^{-1};$
end do;
$sum(\pi_n \cdot X^n, n = 0..K + 1); \ \pi s := simplify(\%);$
$rths := subs(w_k = \pi s, dw_k = diff(\pi s, X) - m \cdot \pi s, rths);$
$print(w_k = \pi s \cdot e^{m \cdot X});$
end do;
end proc:

3.5 Exercises

3.1. Complete the analysis necessary to verify that $\theta_N(x, \varepsilon)$ in (3.15) is uniformly $O(1)$ for $0 \le x \le 1$ as $\varepsilon \to 0^+$. In particular, using (3.11) and (3.13), show that

$$\sum_{n=0}^{N-1} \varepsilon^n [\varepsilon u_n''(x) + a(x, \varepsilon) u_n'(x) + b(x, \varepsilon) u_n(x)] - c(x, \varepsilon) = O(\varepsilon^N) \quad (3.94)$$

uniformly for $0 \le x \le 1$ and using (3.12) and (3.14), show that

$$\sum_{n=0}^{N} \varepsilon^n [v_n''(x/\varepsilon) + a(x, \varepsilon) v_n'(x/\varepsilon) + b(x, \varepsilon) v_{n-1}(x/\varepsilon)] = O(\varepsilon^N) \quad (3.95)$$

uniformly for $0 \le x \le 1$.

3.2. Our approach to Problem C demonstrates a general method for confirming the uniform validity of singular perturbation calculations. This method is used in [7] and [12] to treat a nonlinear generalization of Problem C, for example, and also in [2] and [4], where the emphasis is on partial differential equation problems, but without the clarity achievable with Corollary 1 and Corollary 2. An application of the method to treat a nonlinear integral equation generalization of Problem C, in which Corollary 2 again plays a prominent role, is given in [10].

In general, suppose $f(x, X, \varepsilon) \in C^\infty([0,1] \times [0,\infty] \times [0,\varepsilon_o])$ for some $\varepsilon_o > 0$ so that by Corollary 1 there exists $u_n(x) \in C^\infty([0,1])$ and $v_n(X) \in C^\infty([0,\infty])$ with $v_n(\infty) = 0$ such that, for any $N \geq 0$,

$$f(x, x/\varepsilon, \varepsilon) = \sum_{n=0}^{N-1} \varepsilon^n [u_n(x) + v_n(x/\varepsilon)] + O(\varepsilon^N) \qquad (3.96)$$

uniformly for $0 \leq x \leq 1$ as $\varepsilon \to 0^+$. If, in addition, $h(x,y,\varepsilon) \in C^\infty([0,1] \times [a,b] \times [0,\varepsilon_o])$ and $a \leq f(x, X, \varepsilon) \leq b$ for all $(x, X, \varepsilon) \in [0,1] \times [0,\infty] \times [0,\varepsilon_o]$, then $g(x, X, \varepsilon) = h(x, f(x, X, \varepsilon), \varepsilon)$ also is in $C^\infty([0,1] \times [0,\infty] \times [0,\varepsilon_o])$ and therefore, for certain $\phi_n(x) \in C^\infty([0,1])$ and $\psi_n(X) \in C^\infty([0,\infty])$ with $\psi_n(\infty) = 0$, we know

$$g(x, x/\varepsilon, \varepsilon) = \sum_{n=0}^{N-1} \varepsilon^n [\phi_n(x) + \psi_n(x/\varepsilon)] + \varepsilon^N \theta_N(x, \varepsilon), \qquad (3.97)$$

where $\theta_N(x, \varepsilon) = O(1)$ uniformly for $0 \leq x \leq 1$ as $\varepsilon \to 0^+$. Also, in particular,

$$\phi_0(x) = h(x, u_0(x), 0), \qquad (3.98)$$

$$\psi_0(X) = h(0, u_0(0) + v_0(X), 0) - h(0, u_0(0), 0), \qquad (3.99)$$

$$\phi_1(x) = u_1(x) h^{[0,1,0]}(x, u_0(x), 0) + h^{[0,0,1]}(x, u_0(0), 0)$$
$$+ c_0 x^{-1}[h^{[0,1,0]}(x, u_0(x), 0) - h^{[0,1,0]}(0, u_0(0), 0)], \qquad (3.100)$$

where $c_0 = v_0^{[-1]}(\infty)$, and

$$\psi_1(X) = v_1(X) h^{[0,1,0]}(0, u_0(0) + v_0(X), 0) + \sum_{n=1}^{4} A_n(X), \qquad (3.101)$$

where

$$A_1(X) = u_0'(0) X [h^{[0,1,0]}(0, u_0(0) + v_0(X), 0)$$
$$- h^{[1,0,0]}(0, u_0(0), 0) - 2c_0 X^{-1} h^{[0,2,0]}(0, u_0(0), 0)], \qquad (3.102)$$

$$A_2(X) = X[h^{[1,0,0]}(0, u_0(0) + v_0(X), 0)$$

$$-h^{[0,1,0]}(0, u_0(0), 0) - c_0 X^{-1} h^{[1,1,0]}(0, u_0(0), 0)], \quad (3.103)$$

$$A_3(X) = [h^{[0,0,1]}(0, u_0(0) + v_0(X), 0) - h^{[0,0,1]}(0, u_0(0), 0)], \quad (3.104)$$

$$A_4(X) = u_1(0)[h^{[0,1,0]}(0, u_0(0) + v_0(X), 0) - h^{[0,1,0]}(0, u_0(0), 0)]. \quad (3.105)$$

3.3. Another way to calculate the terms of (3.2) when the initial conditions $y(0, \varepsilon) = \alpha(\varepsilon)$, $y^{[1,0]}(0, \varepsilon) = \varepsilon^{-1} \gamma(\varepsilon)$ are given is to begin by first calculating the terms of the inner expansion,

$$I_N y(x, \varepsilon) = \sum_{n=0}^{N-1} \varepsilon^n Y_n(X). \quad (3.106)$$

These satisfy $Y_n(0) = \alpha_n$ and $Y_n'(0) = \gamma_n$ and just as noted for Problem A in Exercise 2.2, they are of the form $Y_n(X) = p_n(X) + q_n(X) \exp[-a(0,0)X]$, so $v_n(X) = Y_n(X) - p_n(X)$ for $0 \leq n \leq N - 1$. Once this calculation is done, the terms of $O_N y(x, \varepsilon)$ which satisfy (3.13), can be computed using $u_n(0) = \alpha_n - v_n(0)$. Write a Maple program to do these calculations.

3.4. There also is another way to calculate the terms of (3.2) when the boundary conditions $y(0, \varepsilon) = \alpha(\varepsilon)$, $y(1, \varepsilon) = \beta(\varepsilon)$ are given. The idea is to first solve (3.1) for $y(x, \varepsilon, \kappa)$, subject to the initial conditions $y(0, \varepsilon, \kappa) = \alpha(\epsilon)$ and $y^{[1,0,0]}(0, \varepsilon, \kappa) = \kappa/\varepsilon$. This yields

$$y(x, \varepsilon, \kappa) = \sum_{n=0}^{N-1} \varepsilon^n \left[u_n(x, \kappa) + v_n(x/\varepsilon, \kappa) \right] + O(\varepsilon^N), \quad (3.107)$$

say, and then, to satisfy $y(1, \varepsilon, \kappa) = \beta(\varepsilon)$, we just have to solve the (linear) equation

$$\beta(\varepsilon) = \sum_{n=0}^{N-1} \varepsilon^n u_n(1, \kappa) + O(\varepsilon^N) \quad (3.108)$$

for

$$\kappa = \sum_{n=0}^{N-1} \gamma_n \varepsilon^n + O(\varepsilon^N), \quad (3.109)$$

then substitute into (3.107) and reduce the result to the final form (3.2), altogether a simple task for Maple. Try this out with your own choice of parameters, using your program from the previous exercise. You can check your results using the program at the end of Section 3.1.

3.5. First, suppose $u(t) \in C^\infty([0, 1])$ and

$$F(x, \varepsilon) = \frac{1}{\varepsilon} e^{-(x/\varepsilon)^k} \int_0^x u(t) e^{(t/\varepsilon)^k} \, dt, \qquad (3.110)$$

where $k \geq 2$ is an integer. Let

$$p(t) = \sum_{m=0}^{k-2} u^{[m]}(0) t^m, \quad q(t) = t^{-k+1} [u(t) - p(t)]. \qquad (3.111)$$

Show that

$$F(x, \varepsilon) = \sum_{m=0}^{k-2} \varepsilon^m u^{[m]}(0) P_m(x/\varepsilon) + \frac{1}{k} \varepsilon^{k-1} [q(x) - q(0) e^{-(x/\varepsilon)^k}]$$

$$+ \frac{1}{k} \varepsilon^{k-1} e^{-(x/\varepsilon)^k} \int_0^x q'(t) e^{(t/\varepsilon)^k} \, dt, \qquad (3.112)$$

where

$$P_m(X) = e^{-X^k} \int_0^X T^m e^{T^k} \, dT, \qquad (3.113)$$

and show that, for $0 \leq m \leq k - 2$, $P_m(X) \in C^\infty([0, \infty])$. Since $q'(t) \in C^\infty([0, 1])$, this process can be repeated indefinitely, and hence, for any $N \geq 0$,

$$F(x, \varepsilon) = \sum_{n=0}^{N-1} \varepsilon^n [u_n(x) + v_n(x/\varepsilon)] + O(\varepsilon^N) \qquad (3.114)$$

uniformly as $\varepsilon \to 0^+$ for $0 \leq x \leq 1$, where $u_n(x) \in C^\infty([0, 1])$, $v_n(X) \in C^\infty([0, \infty])$. Also, $u_n(x) = 0$ for $0 \leq n \leq k - 2$ and $v_n(\infty) = 0$ for all $n \geq 0$.

Second, show by the same process that if

$$F(x, \varepsilon) = \frac{1}{\varepsilon} e^{(x/\varepsilon)^k} \int_x^1 u(t) e^{-(t/\varepsilon)^k} \, dt, \qquad (3.115)$$

then there exists $u_n(x) \in C^\infty([0, 1])$, and $v_n(X)$, $w_n(X) \in C^\infty([0, \infty])$ such that

$$F(x, \varepsilon) = \sum_{n=0}^{N-1} \varepsilon^n [u_n(x) + v_n(x/\varepsilon) + w_n((1-x)/\hat{\varepsilon})] + O(\varepsilon^N) \qquad (3.116)$$

uniformly as $\varepsilon \to 0^+$, for $0 \leq x \leq 1$, where $\hat{\varepsilon} = \varepsilon^k$. In particular, if $x = 1 - t$, then $(x/\varepsilon)^k - (1/\varepsilon)^k = -(t/\hat{\varepsilon})\phi(t)$, where $0 < \phi(t) \in C^\infty([0, 1])$, and hence, by Theorem 1,

$$e^{-(x/\varepsilon)^k + (1/\varepsilon)^k} = \sum_{n=0}^{N-1} \hat{\varepsilon}^n \pi_n(t/\hat{\varepsilon}) e^{-(t/\hat{\varepsilon})} + O(\hat{\varepsilon}^N), \qquad (3.117)$$

where $\pi_n(T)$ is a polynomial, so, also, $w_n(X) = o(X^{-\infty})$ as $X \to \infty$.

Finally, suppose $v(T) \in C^\infty([0,\infty])$ and

$$G(x,\varepsilon) = \varepsilon^{-1} e^{-(x/\varepsilon)^k} \int_0^x v(t/\varepsilon) e^{(t/\varepsilon)^k} \, dt, \qquad (3.118)$$

$$H(x,\varepsilon) = \varepsilon^{-1} e^{(x/\varepsilon)^k} \int_x^1 v(t/\varepsilon) e^{-(t/\varepsilon)^k} \, dt. \qquad (3.119)$$

Show that $G(x,\varepsilon) = V(x/\varepsilon)$, where $V(X) \in C^\infty([0,\infty])$, and show there exists $v_n(X), w_n(X) \in C^\infty([0,\infty])$ such that

$$H(x,\varepsilon) = \sum_{n=0}^{N-1} \varepsilon^n [v_n(x/\varepsilon) + w_n((1-x)/\hat{\varepsilon})], \qquad (3.120)$$

where, as above, $\hat{\varepsilon} = \varepsilon^k$.

3.6. Show that if $y(x,\varepsilon)$ satisfies (3.41) when $d(x,\varepsilon) = 0$, and if

$$\phi(x,\varepsilon) = e^{-\frac{1}{2}\int_0^x a(t,\varepsilon)\,dt}, \qquad (3.121)$$

then $\tilde{y}(x,\varepsilon) = y(x,\varepsilon)/\phi(x,\varepsilon)$ satisfies

$$\varepsilon^2 \tilde{y}'' - [xb(x) + \varepsilon^2 \tilde{c}(x,\varepsilon)]\tilde{y} = 0, \qquad (3.122)$$

where $\tilde{c}(x,\varepsilon) \in C^\infty([0,1] \times [0,\varepsilon_o])$. Show further that if $u(x^4) \in C^\infty([0,1])$ and $v(X^4) \in C^\infty([0,\infty])$, then there exists $u_n(x), v_n(X))$ such that $u_n(x^4) \in C^\infty([0,1])$, $v_n(X^4) \in C^\infty([0,\infty])$ and, for any $N \geq 0$,

$$[u(x) + v(x/\varepsilon^{2/3})]\phi(x,\varepsilon) = \sum_{n=0}^{N-1} \varepsilon^{n/6}[u_n(x) + v_n(x/\varepsilon^{2/3})] + O(\varepsilon^{N/6}) \quad (3.123)$$

uniformly as $\varepsilon \to 0^+$ for $0 \leq x \leq 1$. Therefore, since (3.59) is known to hold when $a(x,\varepsilon) = 0$, it also holds when $a(x,\varepsilon) \neq 0$.

3.7. Show that if $y(x,\varepsilon)$ is the solution to

$$\varepsilon^2 y'' - (x + 2x^2 + \varepsilon^2)y = 5 - 3x \qquad (3.124)$$

satisfying $y(0,\varepsilon) = 2$ and $y(1,\varepsilon) = \varepsilon$, then $y(x,\varepsilon)$ is approximately

$$c(x,\varepsilon) = \hat{\varepsilon}^{-1} v_1(x/\hat{\varepsilon}) + [u(x) + v_2(x/\hat{\varepsilon}) + w((1-x)/\varepsilon)], \qquad (3.125)$$

where $\hat{\varepsilon} = \varepsilon^{2/3}$,

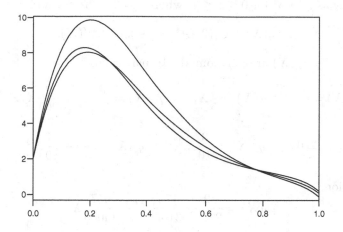

Fig. 3.3 Numerical solution to (3.124) and asymptotic approximations of the solution when $\varepsilon = 0.1$.

$$v_1(X) = \frac{5P(0)}{Ai(0)}Ai(X) - 5P(X), \quad u(x) = \frac{-13}{1+2x}, \quad w(X) = -\frac{2}{3}e^{-\sqrt{3}X}$$

$$\text{(3.126)}$$

and

$$v_2(X) = \frac{11}{Ai(0)}Ai(X) - \frac{2P(0)}{Ai(0)}[XAi(X) - X^2Ai'(X)]$$

$$+2XP(X) - 2X^2P'(X) + 4. \quad \text{(3.127)}$$

Figure 3.3 shows this approximation and Maple's numerical solution for $y(x,\varepsilon)$ when $\varepsilon = 0.1$. It also shows $c(x,\varepsilon)$ plus $\hat{\varepsilon}v_3(x/\hat{\varepsilon})$, the next term after $c(x,\varepsilon)$ in the uniformly valid asymptotic expansion of $y(x,\varepsilon)$, which is significantly closer to the numerical solution.

3.8. The Bessel function $J_\nu(\nu(1-x))$ is the solution to

$$(1-x)^2y'' - (1-x)y' - \nu^2(2x - x^2)y = 0 \quad \text{(3.128)}$$

satisfying $y(0,\nu) = J_\nu(\nu)$ and, for $\nu > 0$, $y(1,\nu) = 0$. Let $\varepsilon = 2^{-1/2}\nu^{-1}$ and use $ProbE$ together with the result of Exercise 1.6 to compute the first several terms of the remarkably simple result (compared, for example, to (9.3.35) of [1], also on page 425 of [6], or the result in [8]),

$$(\nu/2)^{1/3}J_\nu(\nu(1-x)) = \sum_{n=0}^{N-1}\hat{\varepsilon}^n v_n(x/\hat{\varepsilon}) + O(\hat{\varepsilon}^N) \quad \text{(3.129)}$$

uniformly as $\nu \to \infty$ for $0 \le x \le 1$, where $\hat{\varepsilon} = 2^{-1/3}\nu^{-2/3}$ and

$$v_n(X) = p_n(X)Ai(X) + q_n(X)Ai'(X), \qquad (3.130)$$

where $p_n(X)$, $q_n(X)$ are polynomials. In particular,

$$p_0(X) = 1, \qquad p_1(X) = \frac{1}{5}X, \qquad p_2(X) = \frac{3}{35}X^2 + \frac{9}{200}X^5, \qquad (3.131)$$

$$q_0(X) = 0, \qquad q_1(X) = \frac{3}{10}X^2, \qquad q_2(X) = \frac{1}{35} + \frac{17}{70}X^3. \qquad (3.132)$$

In addition,

$$p_3(X) = \frac{-2}{225} + \frac{173}{3150}X^3 + \frac{957}{14000}X^6, \qquad (3.133)$$

$$p_4(X) = \frac{-947}{173250}X + \frac{5903}{138600}X^4 + \frac{23573}{294000}X^7 + \frac{27}{80000}X^{10}, \qquad (3.134)$$

$$q_3(X) = \frac{37}{1575}X + \frac{611}{3150}X^4 + \frac{9}{2000}X^7, \qquad (3.135)$$

$$q_4(X) = \frac{158}{12375}X^2 + \frac{110767}{693000}X^5 + \frac{549}{56000}X^8. \qquad (3.136)$$

3.9. Write a Maple program for Problem D and use it to verify (3.40). Also verify that

$$u_1(x) = \frac{-4x^2}{(2+x)^2}, \qquad v_1(X) = \frac{1}{4}(-5X - 2\sqrt{2}X - \sqrt{2}X^2)e^{-\sqrt{2}X}, \qquad (3.137)$$

$$w_1(X) = \frac{1}{72}(32 - 55X + 20\sqrt{3}X + 5\sqrt{3}X^2)e^{-\sqrt{3}X} \qquad (3.138)$$

for this example.

Chapter 4
Logarithm Problems

4.1 Preliminaries

We saw in Chapter 1 that logarithms arose as a result of integrating a function satisfying the conditions of Theorem 1. Suppose, for a general result, that $f(x, X, \varepsilon) \in C^{\infty}([0,1] \times [0,\infty] \times [0, \varepsilon_o])$ for some $\varepsilon_o > 0$ and

$$y(x, \varepsilon) = x^{-1} \int_0^x f(t, t/\varepsilon, \varepsilon) \, dt. \qquad (4.1)$$

We know, given $N \geq 0$, that

$$f(t, t/\varepsilon, \varepsilon) = \sum_{n=0}^{N-1} \varepsilon^n [u_n(t) + v_n(t/\varepsilon)] + O(\varepsilon^N) \qquad (4.2)$$

uniformly for $0 \leq t \leq 1$ as $\varepsilon \to 0^+$ for certain $u_n(t) \in C^{\infty}([0,1])$ and $v_n(T) \in C^{\infty}([0,\infty])$ with $v_n(\infty) = 0$. If we let

$$U_n(x) = x^{-1} \int_0^x u_n(t) \, dt, \qquad (4.3)$$

and

$$V_n(X) = X^{-1} \int_0^X [v_n(T) - c_n(1+T)^{-1}] \, dT, \qquad (4.4)$$

where $c_n = v_n^{[-1]}(\infty)$, then $U_n(x) \in C^{\infty}([0,1])$ and $V_n(X) \in C^{\infty}([0,\infty])$. Therefore

$$y(x, \varepsilon) = \sum_{n=0}^{N-1} \varepsilon^n [U_n(x) + V_n(x/\varepsilon) + c_n L_1(x/\varepsilon)] + O(\varepsilon^N), \qquad (4.5)$$

uniformly as $\varepsilon \to 0^+$, where $L_1(X) = X^{-1} \ln(1 + X)$. Note that $V_n(\infty) = 0$ and $0 \leq L_1(X) \leq 1$.

Repeated integration gives rise to powers of logarithms. Let $L_0(X) = 1$, $L_n(X) = X^{-1} \ln^n(1 + X)$ for integer $n \geq 1$ and note that $L_n(X) = O(1)$ for $0 \leq X \leq \infty$. If, for example, $v(T) \in C^\infty([0, \infty])$, $v(\infty) = 0$ and

$$F(X) = X^{-1} \int_0^X v(T) \ln(1 + T) \, dT, \tag{4.6}$$

then

$$F(X) = \frac{1}{2} c L_2(X) + X^{-1} \int_0^X \phi(T) \ln(1 + T) \, dT, \tag{4.7}$$

where $c = v^{[-1]}(\infty)$ and $\phi(T) = v(T) - c(1 + T)^{-1} = O(T^{-2})$. Therefore, after an integration by parts,

$$F(X) = \frac{1}{2} c L_2(X) + \phi_1(X) L_1(X) + \phi_0(X), \tag{4.8}$$

where

$$\phi_1(X) = -\int_X^\infty \phi(T) \, dT, \quad \phi_0(X) = -X^{-1} \int_0^X \phi_1(T)(1 + T)^{-1} \, dT. \tag{4.9}$$

By induction, if

$$F(X) = X^{-1} \int_0^X v(T) \ln^n(1 + T) \, dT, \tag{4.10}$$

where n is a positive integer, there exist $\phi_k(X) \in C^\infty([0, \infty])$ for $0 \leq k \leq n+1$ such that

$$F(X) = \sum_{k=0}^{n+1} \phi_k(X) L_k(X). \tag{4.11}$$

In particular $\phi_{n+1}(X) = c/(n + 1)$. Also, $\phi_k(\infty) = 0$ for $0 \leq k \leq n$.

As another example, suppose

$$y(x, \varepsilon) = x^{-1} \int_0^x t^2 u(t, \varepsilon) \ln \frac{1 + t/\varepsilon}{1 + x/\varepsilon} \, dt \tag{4.12}$$

for $\varepsilon > 0$, where $u(x, \varepsilon) \in C^\infty([0, 1] \times [0, \varepsilon_o])$ for some $\varepsilon_o > 0$. If we integrate by parts and define

$$a(t, \varepsilon) = t^{-3} \int_0^t s^2 u(s, \varepsilon) \, ds, \tag{4.13}$$

then $a(t, \varepsilon) \in C^\infty([0, 1] \times [0, \varepsilon_o])$ and, since

$$\frac{t^3}{t + \varepsilon} = t^2 - \varepsilon t + \varepsilon^2 - \frac{\varepsilon^3}{t + \varepsilon}, \tag{4.14}$$

we have

$$y(x, \varepsilon) = x^{-1} \int_0^x (-t^2 + \varepsilon t - \varepsilon^2) a(t, \varepsilon) \, dt + \varepsilon^3 x^{-1} \int_0^x \frac{a(t, \varepsilon)}{t + \varepsilon} \, dt. \quad (4.15)$$

Furthermore, from Exercise 1.7, there exist $g(t, \varepsilon) \in C^\infty([0, 1] \times [0, \varepsilon_o])$ and $c(\varepsilon) \in C^\infty([0, \varepsilon_o])$ such that

$$\frac{a(t, \varepsilon)}{t + \varepsilon} = g(t, \varepsilon) + \frac{c(\varepsilon)}{t + \varepsilon}. \quad (4.16)$$

Therefore

$$y(x, \varepsilon) = x^2 a_0(x, \varepsilon) + \varepsilon x a_1(x, \varepsilon) + \varepsilon^2 a_2(x, \varepsilon) + \varepsilon^2 c(\varepsilon) L_1(x/\varepsilon) \quad (4.17)$$

for all $(x, \varepsilon) \in [0, 1] \times (0, \varepsilon_o]$, where

$$a_0(x, \varepsilon) = -x^{-3} \int_0^x t^2 a(t, \varepsilon) \, dt, \quad a_1(x, \varepsilon) = x^{-2} \int_0^x t a(t, \varepsilon) \, dt, \quad (4.18)$$

$$a_2(x, \varepsilon) = x^{-1} \int_0^x [-a(t, \varepsilon) + \varepsilon g(t, \varepsilon)] \, dt. \quad (4.19)$$

In fact (see Exercise 4.1), for any integers $n \geq 0$ and $m \geq 1$, if

$$y(x, \varepsilon) = x^{-1} \int_0^x t^n u(t, \varepsilon) \ln^m \frac{1 + t/\varepsilon}{1 + x/\varepsilon} \, dt \quad (4.20)$$

for $\varepsilon > 0$, where $u(x, \varepsilon) \in C^\infty([0, 1] \times [0, \varepsilon_o])$, then there exists $a_k(x, \varepsilon) \in C^\infty([0, 1] \times [0, \varepsilon_o])$ for $0 \leq k \leq n$ and $c_k(\varepsilon) \in C^\infty([0, \varepsilon_o])$ for $1 \leq k \leq m$ such that

$$y(x, \varepsilon) = \sum_{k=0}^{n} \varepsilon^k x^{n-k} a_k(x, \varepsilon) + \varepsilon^n \sum_{k=1}^{m} c_k(\varepsilon) L_k(x/\varepsilon) \quad (4.21)$$

for all $(x, \varepsilon) \in [0, 1] \times (0, \varepsilon_o]$. We will use this result to help establish the asymptotic form of the solution to Problem F in the next section.

Another source of logarithms is exponentiation. For example, if $y(x, \varepsilon) = 1 + x/\varepsilon$ for $0 \leq x \leq 1$ and $\varepsilon > 0$, then

$$y^\varepsilon(x, \varepsilon) = e^{\varepsilon \ln(1 + x/\varepsilon)} = 1 + x \sum_{n=0}^{N-1} \frac{1}{(n+1)!} \varepsilon^n L_{n+1}(x/\varepsilon) + O(\varepsilon^N) \quad (4.22)$$

uniformly for $0 \leq x \leq 1$ as $\varepsilon \to 0^+$. Taking this a step further, if $d(\varepsilon) \in C^\infty([0, \varepsilon_o])$, then there exists constants d_{nk} such that

$$y^{\varepsilon d(\varepsilon)}(x, \varepsilon) = 1 + x \sum_{n=0}^{N-1} \varepsilon^n \sum_{k=0}^{n} d_{nk} L_{k+1}(x/\varepsilon) + O(\varepsilon^N) \quad (4.23)$$

uniformly for $0 \le x \le 1$ as $\varepsilon \to 0^+$. In particular,

$$d_{00} = d_0, \quad d_{10} = d_1, \quad d_{11} = \frac{1}{2}d_0^2, \quad d_{20} = d_2,$$

$$d_{21} = d_0 d_1 \quad d_{22} = \frac{1}{6}d_0^3, \tag{4.24}$$

where $d_n = d^{[n]}(0)$. Also, by combining (4.2) and (4.22), observe that if

$$y(x,\varepsilon) = (1 + x/\varepsilon)^\varepsilon f(x, x/\varepsilon, \varepsilon), \tag{4.25}$$

where $f(x, X, \varepsilon) \in C^\infty([0,1] \times [0,\infty] \times [0,\varepsilon_o])$, then there exists $u_{nm}(x) \in C^\infty([0,1])$ and $v_{nm}(X) \in C^\infty([0,\infty])$ such that

$$y(x,\varepsilon) = \sum_{n=0}^{N-1} \varepsilon^n \sum_{m=0}^{n+1} [u_{nm}(x) + v_{nm}(x/\varepsilon)]L_m(x/\varepsilon) + O(\varepsilon^N) \tag{4.26}$$

uniformly for $0 \le x \le 1$ as $\varepsilon \to 0^+$. For example, if

$$y(x,\varepsilon) = (3 + \varepsilon + 3x + x^2 + x/\varepsilon)^\varepsilon, \tag{4.27}$$

then (4.26) holds with

$$u_{00}(x) = 1, \quad u_{01}(x) = x, \tag{4.28}$$

$$u_{10}(x) = 0, \quad u_{11}(x) = 0, \quad u_{12}(x) = \frac{1}{2}x, \tag{4.29}$$

$$u_{20}(x) = 3 + x, \quad u_{21}(x) = 2 + 3x + x^2, \quad u_{22}(x) = 0, \quad u_{23}(x) = \frac{1}{6}x, \tag{4.30}$$

and

$$v_{00}(X) = 0, \quad v_{01}(X) = 0, \tag{4.31}$$

$$v_{10}(X) = \ln\frac{3 + X}{1 + X}, \quad v_{11} = 0, \quad v_{12}(X) = 0, \tag{4.32}$$

$$v_{20}(X) \doteq \frac{-8}{3 + X} + \ln^2\frac{3 + X}{1 + X}, \tag{4.33}$$

$$v_{21} = -2 + X\ln\frac{3 + X}{1 + X}, \quad v_{22}(X) = 0, \quad v_{23}(X) = 0. \tag{4.34}$$

4.2 Problem F

Assume $a(x,\varepsilon)$, $b(x,\varepsilon) \in C^\infty([0,1] \times [0,\varepsilon_o])$ for some $\varepsilon_o > 0$. In addition, assume $a(0,0) = I$, a positive integer. In this section we seek an asymptotic expansion for the solution to

$$(x + \varepsilon)y' + a(x, \varepsilon)y = b(x, \varepsilon) \tag{4.35}$$

satisfying $y(0, \varepsilon) = 0$.

We know there exists $d(\varepsilon) \in C^\infty([0, \varepsilon_o])$ and $g(x, \varepsilon) \in C^\infty([0, 1] \times [0, \varepsilon_o])$ such that

$$\frac{a(x, \varepsilon)}{x + \varepsilon} = g(x, \varepsilon) + \frac{I + \varepsilon d(\varepsilon)}{x + \varepsilon} \tag{4.36}$$

for $\varepsilon > 0$. Therefore

$$y(x, \varepsilon) = \frac{h(x, \varepsilon)}{\varepsilon(1 + x/\varepsilon)^I} \int_0^x (1 + t/\varepsilon)^{I-1} \left(\frac{1 + t/\varepsilon}{1 + x/\varepsilon}\right)^{\varepsilon d(\varepsilon)} \psi(t, \varepsilon)\, dt, \tag{4.37}$$

where

$$h(x, \varepsilon) = \exp\left(-\int_0^x g(t, \varepsilon)\, dt\right), \qquad \psi(x, \varepsilon) = b(x, \varepsilon)/h(x, \varepsilon). \tag{4.38}$$

Also, as in (4.23), there exists constants d_{nm}, such that

$$\left(\frac{1 + t/\varepsilon}{1 + x/\varepsilon}\right)^{\varepsilon d(\varepsilon)} = 1 + \sum_{n=1}^N \varepsilon^n \sum_{m=1}^n d_{nm} \ln^m \frac{1 + t/\varepsilon}{1 + x/\varepsilon} + O(\varepsilon^N) \tag{4.39}$$

for $0 \le t \le x \le 1$ and $\varepsilon \to 0^+$.

Let

$$w_m(x, \varepsilon) = \frac{h(x, \varepsilon)}{\varepsilon(1 + x/\varepsilon)^I} \int_0^x (1 + t/\varepsilon)^{I-1} \psi(t, \varepsilon) \ln^m \frac{1 + t/\varepsilon}{1 + x/\varepsilon}\, dt. \tag{4.40}$$

Since

$$x^{-1} \int_0^x (t/\varepsilon)^k \psi(t, \varepsilon)\, dt = (x/\varepsilon)^k \psi_k(x, \varepsilon), \tag{4.41}$$

where $\psi_k(x, \varepsilon) \in C^\infty([0, 1] \times [0, \varepsilon_o])$, we have $w_0(x, \varepsilon) = f_{00}(x, x/\varepsilon, \varepsilon)$, where

$$f_{00}(x, X, \varepsilon) = \frac{Xh(x, \varepsilon)}{(1 + X)^I} \sum_{k=0}^{I-1} \binom{I-1}{k} X^k \psi_k(x, \varepsilon), \tag{4.42}$$

and clearly $f_{00}(x, X, \varepsilon) \in C^\infty([0, 1] \times [0, \infty] \times [0, \varepsilon_o])$. Similarly, for $m \ge 1$, in view of the expression (4.21) for the integral (4.20), there exist $\phi_k(x, \varepsilon) \in C^\infty([0, 1] \times [0, \varepsilon_o])$ and $c_k(\varepsilon) \in C^\infty([0, \varepsilon_o])$ such that

$$w_m(x, \varepsilon) = \frac{(x/\varepsilon)h(x, \varepsilon)}{(1 + x/\varepsilon)^I} \left(\sum_{k=0}^{I-1} (x/\varepsilon)^k \phi_k(x, \varepsilon) + \sum_{k=1}^m c_k(\varepsilon)L_k(x/\varepsilon)\right). \tag{4.43}$$

Hence, for every $m \ge 0$, and $0 \le k \le m$, there exists $f_{mk}(x, X, \varepsilon) \in C^\infty([0, 1] \times [0, \infty] \times [0, \varepsilon_o])$ such that

$$w_m(x,\varepsilon) = \sum_{k=0}^{m} f_{mk}(x, x/\varepsilon, \varepsilon) L_k(x/\varepsilon). \qquad (4.44)$$

Moreover, when we substitute this into

$$y(x,\varepsilon) = w_0(x,\varepsilon) + \sum_{n=0}^{N-1} \varepsilon^n \sum_{m=1}^{n} d_{mn} w_m(x,\varepsilon) + O(\varepsilon^N), \qquad (4.45)$$

it is clear there exists $\phi_{nk}(x, X, \varepsilon) \in C^\infty([0,1] \times [0,\infty] \times [0,\varepsilon_o])$, just linear combinations of the $f_{mk}(x, X, \varepsilon)$, such that

$$y(x,\varepsilon) = \sum_{n=0}^{N-1} \varepsilon^n \sum_{k=0}^{n} \phi_{nk}(x, x/\varepsilon, \varepsilon) L_k(x/\varepsilon) + O(\varepsilon^N). \qquad (4.46)$$

Therefore, envisioning the expansion of each $\phi_{nk}(x, x/\varepsilon, \varepsilon)$ according to Corollary 1, we see, finally, there exists $u_{nm}(x) \in C^\infty([0,1])$ and $v_{nm}(X) \in C^\infty([0,\infty])$, with $v_{nm}(\infty) = 0$, such that

$$y(x,\varepsilon) = \sum_{n=0}^{N-1} \varepsilon^n \sum_{m=0}^{n} [u_{nm}(x) + v_{nm}(x/\varepsilon)] L_m(x/\varepsilon) + O(\varepsilon^N) \qquad (4.47)$$

uniformly as $\varepsilon \to 0^+$ for $0 \le x \le 1$.

Of course, we want to calculate the terms of (4.47) from the differential equation (4.35). It is apparent from (4.47) that $y(x,\varepsilon)$ has an inner expansion of the familiar form,

$$I_N y(x,\varepsilon) = \sum_{n=0}^{N-1} \varepsilon^n Y_n(X), \qquad (4.48)$$

such that $Y(X,\varepsilon) = y(\varepsilon X, \varepsilon) = I_N y(x,\varepsilon) + O(\varepsilon^N)$ uniformly on any finite interval $0 \le X \le \Delta < \infty$ as $\varepsilon \to 0^+$. Indeed, in somewhat tangled form,

$$I_N y(x,\varepsilon) = \sum_{n=0}^{N-1} \varepsilon^n \sum_{m=0}^{n} \left(\sum_{k=0}^{N-1-n} (\varepsilon X)^k u_{nmk} + v_{nm}(X) \right) L_m(X), \qquad (4.49)$$

where $u_{nmk} = u_{nm}^{[k]}(0)$. In untangled form, we find, with help from Maple,

$$Y_n(X) = \sum_{m=0}^{n} Z_{nm}(X) L_m(X), \qquad (4.50)$$

where,

$$Z_{00}(X) = u_{000} + v_{00}(X), \qquad (4.51)$$

$$Z_{10}(X) = u_{100} + u_{001} X + v_{10}(X), \qquad (4.52)$$

$$Z_{11}(X) = u_{110} + v_{11}(X), \tag{4.53}$$

$$Z_{20}(X) = u_{200} + u_{101}X + u_{002}X^2 + v_{20}(X), \tag{4.54}$$

$$Z_{21}(X) = u_{210} + u_{111}X + v_{21}(X), \tag{4.55}$$

$$Z_{22}(X) = u_{220} + v_{22}(X), \tag{4.56}$$

$$Z_{30}(X) = u_{300} + u_{201}X + u_{102}X^2 + u_{003}X^3 + v_{30}(X) \tag{4.57}$$

$$Z_{31}(X) = u_{310} + u_{211}X + u_{112}X^2 + v_{31}(X), \tag{4.58}$$

$$Z_{32}(X) = u_{320} + u_{221}X + v_{32}(X), \tag{4.59}$$

$$Z_{33}(X) = u_{330} + v_{33}(X). \tag{4.60}$$

In general, $Z_{nm}(X) = r_{nm}(X) + v_{nm}(X)$ and

$$r_{nm}(X) = \sum_{k=0}^{n-m} r_{nmk}X^k, \tag{4.61}$$

where $r_{nmk} = u_{n-k,m,k}$. Therefore

$$u_{nm}(x) = \sum_{k=0}^{N-1-n} r_{n+k,m,k}x^k + O(x^{N-n}). \tag{4.62}$$

In addition, (4.47) implies the existence of an outer expansion for $y(x,\varepsilon)$, having the form

$$O_N y(x,\varepsilon) = z_{00}(x) + \sum_{n=1}^{N-1} \varepsilon^n \sum_{m=0}^{n-1} \ln^m(\varepsilon) z_{nm}(x), \tag{4.63}$$

such that $y(x,\varepsilon) = O_N y(x,\varepsilon) + O(\varepsilon^N \ln^{N-1}(\varepsilon))$ uniformly on $0 < \delta \le x \le 1$ as $\varepsilon \to 0^+$. Indeed, in particular, using $v_{nmk} = v_{nm}^{[-k]}(\infty)$, we have

$$z_{00}(x) = u_{00}(x), \tag{4.64}$$

$$z_{10}(x) = v_{001}x^{-1} + u_{10}(x), \tag{4.65}$$

$$z_{20}(x) = v_{002}x^{-2} + v_{101}x^{-1} + x^{-1}u_{11}(x)\ln(x) + u_{20}(x), \tag{4.66}$$

$$z_{21}(x) = -x^{-1}u_{11}(x), \tag{4.67}$$

$$z_{30}(x) = v_{003}x^{-3} + v_{102}x^{-2} + v_{111}x^{-2}\ln(x) + u_{11}(x)x^{-2}$$

$$+v_{201}x^{-1} + u_{21}(x)x^{-1}\ln(x) + u_{22}(x)x^{-1}\ln^2(x) + u_{30}(x), \tag{4.68}$$

$$z_{31}(x) = -x^{-2}v_{111} - u_{21}(x) - 2u_{22}(x)x^{-1}\ln(x), \tag{4.69}$$

$$z_{32}(x) = x^{-1}u_{22}(x). \tag{4.70}$$

The first part of the Maple program *ProbF*, given below, calculates $Y_k(X)$ for $0 \leq k \leq N - 1$, once $a(x, \varepsilon)$, $b(x, \varepsilon)$ and N are prescribed, in the usual way from the differential equation for $Y(X, \varepsilon)$. Then each $Z_{ij}(X)$ for $0 \leq i \leq N - 1$ and $0 \leq j \leq i$ is isolated and split into $r_{ij}(X)$ and $v_{ij}(X)$. Note that this involves a procedure for expanding $Z_{ij}(X)$ as $X \to \infty$ out to precisely $O(X^{-N})$. The resulting series expansions for $u_{ij}(x)$ as $x \to 0$ and $v_{ij}(X)$ as $X \to \infty$ are used in the outer expansion calculations. For example, if we take $a(x, \varepsilon) = 2 + x$, $b(x, \varepsilon) = 1$ and $N = 4$, *ProbF* yields

$$v_{00}(X) = \frac{-1}{2(1 + X)^2}, \qquad v_{10}(X) = \frac{6X - 1}{12(1 + X)^2}, \qquad (4.71)$$

$$v_{11}(X) = \frac{-X}{2(1 + X)^2}, \qquad v_{20}(X) = \frac{21X + 8}{36(1 + X)^2}, \qquad (4.72)$$

$$v_{21}(X) = \frac{-13X - 6}{12(1 + X)^2}, \qquad v_{22}(X) = \frac{-X}{4(1 + X)^2}, \qquad (4.73)$$

and in accordance with (4.62), *ProbF* also yields

$$u_{00}(x) = \frac{1}{2} - \frac{1}{6}x + \frac{1}{24}x^2 - \frac{1}{120}x^3 + O(x^4), \qquad u_{11}(x) = O(x^3), \quad (4.74)$$

$$u_{10}(x) = \frac{1}{12} - \frac{1}{18}x + \frac{29}{144}x^2 + O(x^3), \qquad u_{22}(x) = O(x^2), \qquad (4.75)$$

$$u_{21}(x) = \frac{1}{2} - \frac{1}{4}x + O(x^2), \qquad u_{20}(x) = \frac{-2}{9} + \frac{131}{2160}x + O(x^2). \quad (4.76)$$

In addition, the *ProbF* substitution of (4.63) into (4.35), still with $a(x, \varepsilon) = 2 + x$, $b(x, \varepsilon) = 1$ and $N = 4$, yields the sequence of differential equations

$$xz'_{00} + (2 + x)z_{00} = 1, \qquad xz'_{10} + (2 + x)z_{10} = -z'_{00}(x), \qquad (4.77)$$

$$xz'_{20} + (2 + x)z_{20} = -z'_{10}(x), \qquad xz'_{21} + (2 + x)z_{21} = 0, \qquad (4.78)$$

$$xz'_{31} + (2 + x)z_{31} = -z'_{21}(x), \qquad xz'_{32} + (2 + x)z_{32} = 0, \qquad (4.79)$$

and from the last lines of *ProbF* we get

$$z_{00}(x) = u_{00}(x), \qquad z_{10}(x) = u_{10}(x) \qquad (4.80)$$

$$z_{20}(x) = \frac{1}{2}x^{-1} + x^{-1}\ln(x)u_{11}(x) + u_{20}(x), \quad z_{21}(x) = -x^{-1}u_{11}(x), \quad (4.81)$$

$$z_{31}(x) = \frac{1}{2}x^{-2} - x^{-1}u_{21}(x) - 2x^{-1}\ln(x)u_{22}(x), \qquad (4.82)$$

$$u_{32}(x) = x^{-1}u_{22}(x), \qquad (4.83)$$

in accordance with (4.64)-(4.70).

To complete the calculations for this example, from (4.80a) and (4.74a) we see $z_{00}(0) = 1/2$ and therefore the solution to (4.77a) is

$$u_{00}(x) = \frac{1}{x^2}(e^{-x} - 1 - x) \tag{4.84}$$

Next, the solution to (4.77b) such that $z_{10}(0) = u_{10}(0) = 1/12$ is

$$u_{10}(x) = -\frac{e^{-x}}{x^2}[E(x) + \frac{2}{x}(e^x - 1 - x)], \tag{4.85}$$

where $E(x) = \int_0^x t^{-1}(e^t - 1)\, dt$. In addition, we readily find

$$u_{11}(x) = 0, \qquad u_{22}(x) = 0, \qquad u_{21}(x) = \frac{1}{2x}(1 - e^x), \tag{4.86}$$

and finally, the (convergent) series solution to (4.78a) yields

$$u_{20}(x) = -\frac{2}{9} + \frac{131}{2160}x - \frac{977}{8640}x^2 + \frac{4367}{302400}x^3 + O(x^4). \tag{4.87}$$

Thus we have computed all the terms of (4.47) with $N = 3$, assuming $a(x, \varepsilon) = 2 + x$, $b(x, \varepsilon) = 1$.

```
ProbF := proc(a, b, N)
A := subs(x = ε · X, a); B := subs(x = ε · X, b);
INy := sum(εⁿ · Yₙ, n = 0..N − 1);
INdy := sum(εⁿ · dYₙ, n = 0..N − 1);
INde := series((1 + X) · INdy + A · INy − B, ε = 0, N);
INde := convert(%, polynom);
for k from 0 to N − 1 do
temp := coeff(INde, ε, k);
de := subs(Yₖ = z(X), dYₖ = diff(z(X), X), temp) = 0;
dsolve({de, z(0) = 0});
Yk := rhs(%); dYk := diff(Yk, X);
INde := subs(Yₖ = Yk, dYₖ = dYk, INde);
Yₖ := Yk; end do;
for i from 0 to N − 1 do
Z_{i,0} := coeff(Yᵢ, ln(1 + X), 0);
for j from 1 to i do
Z_{i,j} := X · coeff(Yᵢ, ln(1 + X), j);
end do; end do;
for i from 0 to N − 1 do
for j from 0 to i do
if Z_{i,j} = 0 then sZij := 0 else
temp := 0; k := N − 1;
while order(temp) < N do
```

$k := k + 1;\ sZij := series(Z_{i,j}, X = \infty, k);$
$temp := series(subs(X = X^{-1}, \%), X = 0, k);$
end do; end if;
$r_{i,j} := convert(series(sZij, X = \infty, 1), polynom);$
$vij := simplify(Z_{i,j} - r_{i,j});\ print(v_{i,j} = vij);$
$sv_{i,j} := sZij - r_{i,j};$
end do; end do;
for i **from** 0 **to** $N - 1$ **do**
for j **from** 0 **to** i **do**
$uij := sum(x^l \cdot coeff(r_{i+l,j}, X, l), l = 0..N - 1 - i) + O(x^{N-i});$
$print(u_{i,j} = uij);$
end do; end do;
$ONy := z_{0,0} + sum(\varepsilon^n \cdot sum(\ln(\varepsilon)^m \cdot z_{n,m}, m = 0..n - 1), n = 0..N - 1);$
$ONdy := Dz_{0,0} + sum(\varepsilon^n \cdot sum(\ln(\varepsilon)^m \cdot Dz_{n,m}, m = 0..n-1), n = 0..N-1);$
$eq := (x + \varepsilon) \cdot ONdy + a \cdot ONy - b;$
$ONde := series(eq, \varepsilon = 0, N);$
for i **from** 0 **to** $N - 1$ **do;**
for j **from** 0 **to** i **do;**
$dei := coeff(ONde, \varepsilon, i);\ deij := coeff(dei, \ln(\varepsilon), j);\ print(deij = 0);$
end do; end do; $L_0 := 1;$
for i **from** 1 **to** $N - 1$ **do** $L_i := X^{-1} \cdot \ln(1 + X)^i;$ **end do;**
$y := sum(\varepsilon^n \cdot (sum(L_m \cdot (u_{n,m} + sv_{n,m}), m = 0..n)), n = 0..N - 1);$
$ONy := series(subs(X = x \cdot \varepsilon^{-1}, y), \varepsilon = 0, N);$
for n **from** 0 **to** $N - 1$ **do**
$z_n := coeff(ONy, \varepsilon, n);$
end do;
$print(z_{0,0} = z_0);$
for n **from** 1 **to** $N - 1$ **do**
for m **from** 0 **to** $n - 1$ **do**
$lnz_{n,m} := coeff(z_n, \ln(\varepsilon), m);\ print(z_{n,m} = lnz_{n,m});$
end do; end do;
end proc:

4.3 Problem G, Part 1

Problem G is to derive an asymptotic expansion for the solution to the differential equation

$$\varepsilon^2 y'' + [xa(x) + \varepsilon^2 b(x,\varepsilon)]y' + \varepsilon c(x,\varepsilon)y = d(x,\varepsilon) \qquad (4.88)$$

that satisfies the boundary conditions $y(0,\varepsilon) = \alpha(\varepsilon)$, $y(1,\varepsilon) = \beta(\varepsilon)$. We will assume $b(x,\varepsilon)$, $c(x,\varepsilon)$, $d(x,\varepsilon) \in C^\infty([0,1] \times [0,\varepsilon_o])$ and $\alpha(\varepsilon)$, $\beta(\varepsilon) \in C^\infty([0,\varepsilon_o])$ for some $\varepsilon_o > 0$. We also assume $0 < a(x) \in C^\infty([0,1])$ and, without loss of generality, take $a(0) = 2$, $\alpha(\varepsilon) = 0$. First we will obtain asymptotic expansions for two initial value problems, namely,

$$\varepsilon^2 w'' + [xa(x) + \varepsilon^2 b(x, \varepsilon)]w' + \varepsilon c(x, \varepsilon)w = d(x, \varepsilon) \qquad (4.89)$$

subject to $w(0, \varepsilon) = 0$, $w'(0, \varepsilon) = 0$, and

$$\varepsilon^2 z'' + [xa(x) + \varepsilon^2 b(x, \varepsilon)]z' + \varepsilon c(x, \varepsilon)z = 0 \qquad (4.90)$$

subject to $z(0, \varepsilon) = 0$, $z'(0, \varepsilon) = (2/\sqrt{\pi})\varepsilon^{-1}$. In the end we will determine an asymptotic expansion for the function $\kappa(\varepsilon)$ such that

$$y(x, \varepsilon) = w(x, \varepsilon) + \kappa(\varepsilon)z(x, \varepsilon). \qquad (4.91)$$

This approach to Problem G is essentially the same as the one described in Exercise 3.4 for Problem C.

It is reasonable to expect the solution to

$$\varepsilon^2 w_0'' + xa(x)w_0' = d(x, \varepsilon) \qquad (4.92)$$

such that $w_0(0, \varepsilon) = w_0'(0, \varepsilon) = 0$, the problem obtained by neglecting $\varepsilon^2 b(x, \varepsilon)w' + \varepsilon c(x, \varepsilon)w$ in (4.89), uniformly approximates $w(x, \varepsilon)$. An even better approximation should be the solution to the problem obtained by substituting $\varepsilon^2 b(x, \varepsilon)w_0'(x, \varepsilon) + \varepsilon c(x, \varepsilon)w_0(x, \varepsilon)$ for $\varepsilon^2 b(x, \varepsilon)w' + \varepsilon c(x, \varepsilon)w$. We are going to prove that in fact

$$w(x, \varepsilon) = \sum_{n=0}^{\infty} \varepsilon^n w_n(x, \varepsilon), \qquad (4.93)$$

where, for $n \geq 1$,

$$\varepsilon^2 w_n'' + xa(x)w_n' = -\varepsilon b(x, \varepsilon)w_{n-1}'(x, \varepsilon) - c(x, \varepsilon)w_{n-1}(x, \varepsilon). \qquad (4.94)$$

First, there exists $B > 0$ such that $|b(x, \varepsilon)|$, $|c(x, \varepsilon)|$, $|d(x, \varepsilon)| \leq B$ for all $(x, \varepsilon) \in [0, 1] \times [0, \varepsilon_o]$. Therefore $|\varepsilon w_0'(x, \varepsilon)| \leq BF(x, \varepsilon)$ on $[0, 1] \times (0, \varepsilon_o]$, where

$$F(x, \varepsilon) = \varepsilon^{-1} e^{-(x/\varepsilon)^2 \sigma(x)} \int_0^x e^{(t/\varepsilon)^2 \sigma(t)} \, dt, \qquad (4.95)$$

$$\sigma(x) = x^{-2} \int_0^x ta(t) \, dt. \qquad (4.96)$$

Note that $\sigma(x) > 0$ for $0 \leq x \leq 1$. If we let $\tau = t[\sigma(t)]^{1/2}$ and denote the inverse of this transformation by $t = \theta(\tau)$, and let $\xi = x[\sigma(x)]^{1/2}$, then $|\theta'(\tau)| = 2[\sigma(t)]^{1/2}/[t + 2\sigma(t)] \leq M$ for some $M > 0$, so $F(x, \varepsilon) \leq MP(\xi/\varepsilon)$, still on $[0, 1] \times (0, \varepsilon_o]$, where

$$P(Z) = e^{-Z^2} \int_0^Z e^{T^2} \, dT. \tag{4.97}$$

Also, $P(Z) = O(Z)$ as $Z \to 0$ and $P(Z) = O(Z^{-1})$ as $Z \to \infty$. Hence, there exists $K > 0$ such that $P(Z) \le KZ/(1 + Z^2)$ for $0 \le Z \le \infty$. In addition,

$$\frac{\xi/\varepsilon}{1 + (\xi/\varepsilon)^2} = \frac{x/\varepsilon}{1 + (x/\varepsilon)^2} f(x, x/\varepsilon), \tag{4.98}$$

where

$$f(x, X) = \frac{(1 + X^2)[\sigma(x)]^{1/2}}{1 + X^2 \sigma(x)}, \tag{4.99}$$

so $f(x, x/\varepsilon) = O(1)$ on $[0, 1] \times (0, \varepsilon_o]$, by Theorem 1. Therefore $F(x, \varepsilon) \le A(x/\varepsilon)/[1 + (x/\varepsilon)^2]$ for some constant $A > 0$. Furthermore, $X/(1 + X^2) \le 2/(1 + X)$ for $0 \le X \le \infty$. Therefore

$$|w_0(x, \varepsilon)| \le 2AB\varepsilon^{-1} \int_0^x \frac{dt}{1 + t/\varepsilon} \le 2AB \ln(1 + 1/\varepsilon) \tag{4.100}$$

for all $(x, \varepsilon) \in [0, 1] \times (0, \varepsilon_o]$. In addition, $X/(1 + X^2) \le 2 \ln(1 + X)$. Therefore $|\varepsilon w_0'(x, \varepsilon)| \le 2AB \ln(1 + 1/\varepsilon)$, too.

Next, we now know

$$|\varepsilon b(x, \varepsilon) w_0'(x, \varepsilon) + c(x, \varepsilon) w_0(x, \varepsilon)| \le 4AB^2 \ln(1 + 1/\varepsilon) \tag{4.101}$$

for $0 \le x \le 1$ and $0 < \varepsilon \le \varepsilon_o$. Consequently, $|\varepsilon w_1'(x, \varepsilon)| \le 4AB^2 \ln(1 + 1/\varepsilon)F(x, \varepsilon)$, and therefore both $|w_1(x, \varepsilon)|$ and $|\varepsilon w_1'(x, \varepsilon)|$ are bounded by $8A^2B^2 \ln^2(1 + 1/\varepsilon)$. Indeed, by induction, if we let $C = 4AB$, then for any $n \ge 0$, both $|w_n(x, \varepsilon)|$ and $|\varepsilon w_n'(x, \varepsilon|$ are less than or equal to $(1/2)C^{n+1} \ln^{n+1}(1 + 1/\varepsilon)$ for all $(x, \varepsilon) \in [0, 1] \times (0, \varepsilon_o]$. So, of the three sums,

$$\sum_{n=0}^{\infty} \varepsilon^n w_n(x, \varepsilon), \qquad \sum_{n=0}^{\infty} \varepsilon^n w_n'(x, \varepsilon), \qquad \sum_{n=0}^{\infty} \varepsilon^n w_n''(x, \varepsilon), \tag{4.102}$$

we know the first two converge absolutely and uniformly for all $(x, \varepsilon) \in [0, 1] \times (0, \varepsilon_o]$, assuming ε_o is set sufficiently small, and therefore the second sum is the derivative of the first. The third sum also converges absolutely and uniformly, and therefore is the second derivative of the first sum. This is seen by summing both sides of (4.94) times ε^n, and this shows simultaneously that the first sum in (4.102) is indeed the desired solution of (4.89).

To ascertain the asymptotic form of $w(x, \varepsilon)$, we need to examine the terms of (4.93) in more detail. Let $\lambda_m(x, \varepsilon) = xL_m(x/\varepsilon)$ for $m \ge 1$ and $\lambda_0(x, \varepsilon) = 1$.

Proposition 1. *Assume* $f(x, X, \varepsilon) \in C^\infty([0, 1] \times [0, \infty] \times [0, \varepsilon_o])$ *and let*

$$g_m(x, \varepsilon) = \varepsilon^{-1} e^{-(x/\varepsilon)^2 \sigma(x)} \int_0^x e^{(t/\varepsilon)^2 \sigma(t)} f(t, t/\varepsilon, \varepsilon) \lambda_m(t, \varepsilon) \, dt, \tag{4.103}$$

where $0 < \sigma(x) \in C^\infty([0,1])$. *There exists* $\phi_{mnk}(x) \in C^\infty([0,1])$ *and* $\psi_{mnk}(X) \in C^\infty([0,\infty])$ *with* $\psi_{mnk}(\infty) = 0$ *such that for any* $N \geq 0$,

$$g_m(x,\varepsilon) = \sum_{n=0}^{N-1} \varepsilon^n \sum_{k=0}^{m} [\phi_{mnk}(x) + \psi_{mnk}(x/\varepsilon)]\lambda_k(x,\varepsilon) + O(\varepsilon^N) \cdot \quad (4.104)$$

uniformly as $\varepsilon \to 0^+$ *for* $0 \leq x \leq 1$. *In particular,* $\phi_{m,0,k}(x) = 0$ *for* $0 \leq k \leq m$ *and, for* $m \geq 1$, $\psi_{m,0,0}(X) = \phi_{m,1,0}(x) = 0$.

Proof. We know this is true for $m = 0$ from our work on Problem E. For $m \geq 1$, we readily get (4.104) from (4.103) using integration by parts. In particular,

$$g_1(x,\varepsilon) = g_0(x,\varepsilon)\lambda_1(x,\varepsilon) - \varepsilon\hat{g}_0(x,\varepsilon), \quad (4.105)$$

where $\hat{g}_0(x,\varepsilon)$ is the same as $g_0(x,\varepsilon)$, except $g_0(t,\varepsilon)/(1+t/\varepsilon)$ occurs in place of $f(t,t/\varepsilon,\varepsilon)$. Therefore (4.104) holds for $m = 1$ with $\phi_{1,0,1}(x) = 0$ and $\phi_{1,0,0}(x) = \psi_{1,0,0}(X) = \phi_{1,1,0}(x) = 0$. For $m \geq 2$,

$$g_m(x,\varepsilon) = g_0(x,\varepsilon)\lambda_m(x,\varepsilon) - \hat{g}_{m-1}(x,\varepsilon), \quad (4.106)$$

where $\hat{g}_{m-1}(x,\varepsilon)$ is the same as $g_{m-1}(x,\varepsilon)$, except $mg_0(t,\varepsilon)/(1 + t/\varepsilon)$ occurs in place of $f(t,t/\varepsilon,\varepsilon)$. Therefore $\phi_{m,0,m}(x) = 0$ and if we assume $\phi_{m-1,0,k}(x) = 0$ for $0 \leq k \leq m - 1$, $\psi_{m-1,0,0}(X) = 0$ and $\phi_{m-1,1,0}(x) = 0$, then $\phi_{m,0,k}(x) = 0$ for $0 \leq k \leq m - 1$, $\psi_{m,0,0}(X) = 0$ and $\phi_{m,1,0}(x) = 0$.

It follows from Proposition 1 with $m = 0$ that

$$\varepsilon w_0'(x,\varepsilon) = \sum_{n=0}^{N-1} \varepsilon^n [u_{0,n,0}(x) + v_{0,n,0}(x/\varepsilon)] + O(\varepsilon^N), \quad (4.107)$$

where $u_{0,n,0}(x) \in C^\infty([0,1])$, $v_{0,n,0}(X) \in C^\infty([0,\infty])$ and $u_{0,0,0}(x) = 0$. Therefore, as in the first paragraph of Section 4.1,

$$\varepsilon w_0(x,\varepsilon) = \sum_{n=0}^{N-1} \varepsilon^n \sum_{k=0}^{1} [U_{0,n,k}(x) + V_{0,n,k}(x/\varepsilon)]\lambda_k(x,\varepsilon) + O(\varepsilon^N) \quad (4.108)$$

uniformly as $\varepsilon \to 0^+$ for $0 \leq x \leq 1$, where $U_{0,n,0}(x) \in C^\infty([0,1])$, $V_{0,n,0}(X) \in C^\infty([0,\infty])$ and, in particular, $U_{0,0,0}(x) = V_{0,0,0}(X) = 0$ and $U_{0,n,1}(x) + V_{0,n,1}(x/\varepsilon) = c_{n,1}$, a constant. To proceed further, we need a more general result.

Proposition 2. *Assume* $f(x,X,\varepsilon) \in C^\infty([0,1] \times [0,\infty] \times [0,\varepsilon_o])$ *for some* $\varepsilon_o > 0$. *Assume also that* $f(x,\infty,0) = 0$ *and let*

$$g_m(x,\varepsilon) = \int_0^x f(t,t/\varepsilon,\varepsilon) \ln^m(1 + t/\varepsilon) \, dt. \quad (4.109)$$

There exists $\phi_{mnk}(x) \in C^\infty([0,1])$ and $\psi_{mnk}(X) \in C^\infty([0,\infty])$ such that

$$g_m(x,\varepsilon) = \sum_{n=0}^{N-1} \varepsilon^n \sum_{k=0}^{m+1} [\phi_{mnk}(x) + \psi_{mnk}(x/\varepsilon)]\lambda_k(x,\varepsilon) + O(\varepsilon^N) \qquad (4.110)$$

uniformly as $\varepsilon \to 0^+$ for $0 \le x \le 1$. In particular, $\phi_{m,0,0}(x) = \psi_{m,0,0}(X) = 0$ and for each $n \ge 0$, $\phi_{m,n,m+1}(x) + \psi_{m,n,m+1}(x/\varepsilon) = c_{n,m+1}$, a constant.

Proof. This follows from the second paragraph of Section 4.1, in particular, from our analysis of (4.20) with $n = 0$, or, independently, using integration by parts. We leave the details as an exercise.

At this point, from (4.107) and (4.108), and Corollary 1, we can see

$$\varepsilon[\varepsilon b(x,\varepsilon)w_0'(x,\varepsilon) + c(x,\varepsilon)w_0(x,\varepsilon)] =$$
$$\sum_{n=0}^{N-1} \varepsilon^n \sum_{k=0}^{1} [\phi_{nk}(x) + \psi_{nk}(x/\varepsilon)]\lambda_k(x,\varepsilon) + O(\varepsilon^N), \qquad (4.111)$$

where $\phi_{nk}(x) \in C^\infty([0,1])$, $\psi_{nk}(X) \in C^\infty([0,\infty])$ and $\phi_{0,0}(x) = \psi_{0,0}(X) = 0$. Therefore, by Proposition 1, there exists $u_{1,n,k}(x) \in C^\infty([0,1])$ and $v_{1,n,k}(X) \in C^\infty([0,\infty])$ such that, after canceling an ε factor,

$$\varepsilon w_1'(x,\varepsilon) = \sum_{n=0}^{N-1} \varepsilon^n \sum_{k=0}^{1} [u_{1,n,k}(x) + v_{1,n,k}(x/\varepsilon)] \ln^k(1+x/\varepsilon) + O(\varepsilon^N). \quad (4.112)$$

In particular, $u_{1,0,k}(x) = 0$ for $k = 0, 1$. Thus we can apply Proposition 2 to this result, and therefore

$$\varepsilon w_1(x,\varepsilon) = \sum_{n=0}^{N-1} \varepsilon^n \sum_{k=0}^{2} [U_{1,n,k}(x) + V_{1,n,k}(x/\varepsilon)]\lambda_k(x,\varepsilon) + O(\varepsilon^N), \quad (4.113)$$

where, in particular, $U_{1,0,0}(x) = V_{1,0,0}(X) = 0$. Obviously, we can iterate this process indefinitely. Hence we conclude, there exists $U_{nmk}(x) \in C^\infty([0,1])$ and $V_{nmk}(X) \in C^\infty([0,\infty])$ such that for every $m \ge 0$,

$$\varepsilon w_m(x,\varepsilon) = \sum_{n=0}^{N-1} \varepsilon^n \sum_{k=0}^{m+1} [U_{mnk}(x) + V_{nmk}(x/\varepsilon)]\lambda_k(x,\varepsilon) \qquad (4.114)$$

uniformly as $\varepsilon \to 0^+$ for $0 \le x \le 1$. It is also true, by virtually the same reasoning, that $z(x,\varepsilon)$, our solution to (4.90), has an infinite series solution,

$$z(x,\varepsilon) = \sum_{m=0}^{\infty} \varepsilon^m z_m(x,\varepsilon), \qquad (4.115)$$

and $\varepsilon z_m(x, \varepsilon)$ has an asymptotic expansion of the same form as $\varepsilon w_m(x, \varepsilon)$. See Exercise 4.5.

4.4 Problem G, Part 2

We can see upon substituting (4.114) into (4.93) that if we let $\widetilde{w}(x, \varepsilon) = \varepsilon^2 w(x, \varepsilon)$, then there exists $u_{nm}(x) \in C^\infty([0, 1])$ and $v_{nm}(X) \in C^\infty([0, \infty])$ such that, for any $N \geq 0$,

$$\widetilde{w}(x, \varepsilon) = \sum_{n=0}^{N-1} \varepsilon^n \sum_{m=0}^{n} [u_{nm}(x) + v_{nm}(x/\varepsilon)] L_m(x/\varepsilon) + O(\varepsilon^N) \qquad (4.116)$$

uniformly as $\varepsilon \to 0^+$ for $0 \leq x \leq 1$. That is, $\widetilde{w}(x, \varepsilon)$ has the same asymptotic form as the solution to Problem F. Hence, if we denote the N-term expansion of $\widetilde{W}(X, \varepsilon) = \widetilde{w}(\varepsilon X, \varepsilon)$ by

$$I_N \widetilde{w}(x, \varepsilon) = \sum_{n=0}^{N-1} \varepsilon^n \widetilde{W}_n(X), \qquad (4.117)$$

then, as with $Y_n(X)$ in Section 4.2,

$$\widetilde{W}_n(X) = \sum_{m=0}^{n} Z_{nm}(X) L_m(X) \qquad (4.118)$$

and for the leading terms we have (4.51)-(4.60). Likewise, for the N-term outer expansion

$$O_N \widetilde{w}(x, \varepsilon) = z_{00}(x) + \sum_{n=1}^{N-1} \varepsilon^n \sum_{m=0}^{n-1} \ln^m(\varepsilon) z_{nm}(x), \qquad (4.119)$$

we have (4.64)-(4.70). The differential equation for $\widetilde{w}(x, \varepsilon)$ is

$$\varepsilon^2 \widetilde{w}'' + [xa(x) + \varepsilon^2 b(x, \varepsilon)] \widetilde{w}' + \varepsilon c(x, \varepsilon) \widetilde{w} = \varepsilon^2 d(x, \varepsilon), \qquad (4.120)$$

of course, and it is readily determined that

$$xa(x) z_{00}' = 0, \qquad (4.121)$$

$$xa(x) z_{10}' = -c_0(x) z_{00}(x), \qquad (4.122)$$

$$xa(x) z_{21}' = 0 \qquad (4.123)$$

$$xa(x) z_{20}' = -b_0(x) z_{00}'(x) - z_{00}''(x) - c_0(x) z_{10}(x) - c_1(x) z_{20}(x) + d_0(x), \qquad (4.124)$$

$$xa(x)z'_{32} = 0, \tag{4.125}$$

$$xa(x)z'_{31} = -c_0(x)z_{21}(x), \tag{4.126}$$

where $b_0(x) = b(x,0)$, $c_0(x) = c(x,0)$, $d_0(x) = d(x,0)$ and $c_1(x) = c^{[0,1]}(x,0)$. From the differential equation for $\widetilde{W}(X,\varepsilon)$, it is apparent that

$$\widetilde{W}''_0 + 2X\widetilde{W}'_0 = 0 \tag{4.127}$$

and therefore, since $\widetilde{W}_n(0) = \widetilde{W}'_n(0) = 0$ for all $n \geq 0$, $\widetilde{W}_0(X) = 0$. From (4.51), first with $X = \infty$, this means

$$u_{00}(0) = 0, \qquad v_{00}(X) = 0. \tag{4.128}$$

It follows also that

$$\widetilde{W}''_1 + 2X\widetilde{W}'_1 = 0, \tag{4.129}$$

so $\widetilde{W}_1(X) = 0$, too, and therefore, by (4.52) and (4.53),

$$u_{10}(0) = u'_{00}(0) = u_{11}(0) = 0, \quad v_{10}(X) = v_{11}(X) = 0. \tag{4.130}$$

Now it is apparent that

$$\widetilde{W}''_2 + 2X\widetilde{W}'_2 = d_{00}, \tag{4.131}$$

where $d_{00} = d(0,0)$. Therefore $\widetilde{W}'_2(X) = d_{00}P(X)$, where $P(X) = \frac{1}{2}X^{-1} + O(X^{-3})$ is the function defined by (4.97). Hence, if we let

$$Q(X) = \int_X^\infty [P(T) - \frac{1}{2}(1+T)^{-1}]\,dT, \tag{4.132}$$

then $Q(X) \in C^\infty([0,\infty])$, in particular, $Q(X) = \frac{1}{2}X^{-1} + O(X^{-2})$ as $X \to \infty$, and

$$\widetilde{W}_2(X) = d_{00}[Q(0) - Q(X)] + \frac{1}{2}d_{00}\ln(1+X). \tag{4.133}$$

It follows that $Z_{20}(X) = d_{00}[Q(0) - Q(X)]$ and therefore

$$u_{20}(0) = d_{00}Q(0), \quad u'_{10}(0) = u''_{00}(0) = 0, \quad v_{20}(X) = -d_{00}Q(X). \tag{4.134}$$

Also, $Z_{21}(X) = \frac{1}{2}d_{00}X$ and $Z_{22}(X) = 0$, so

$$u'_{11}(0) = \frac{1}{2}d_{00}, \qquad u_{21}(0) = u_{22}(0) = 0, \qquad v_{21}(X) = v_{22}(X) = 0. \tag{4.135}$$

Turning to the outer expansion of $\widetilde{w}(x,\varepsilon)$, from (4.121), (4.64) and (4.128a), we see

$$z_{00}(x) = u_{00}(x) = 0. \tag{4.136}$$

Therefore from (4.122), along with (4.65) and (4.128b), we also have

$$z_{10}(x) = u_{10}(x) = 0. \tag{4.137}$$

Similarly, from (4.123), (4.130a) and (4.67), we find

$$z_{21}(x) = -\frac{1}{2}d_{00}, \quad u_{11}(x) = \frac{1}{2}d_{00}x. \tag{4.138}$$

To determine $u_{20}(x)$, we currently have $xa(x)z'_{20} = d_0(x)$ from (4.124) and $z_{20}(x) = \frac{1}{2}d_{00}\ln(x) + u_{20}(x)$ from (4.66). Therefore, noting (4.134a), we find

$$u_{20}(x) = d_{00}Q(0) + D(x), \tag{4.139}$$

where $D(x) \in C^\infty([0,1])$ is given by

$$D(x) = \int_0^x \left[\frac{d_0(t)}{a(t)} - \frac{d_{00}}{2}\right] \frac{dt}{t}. \tag{4.140}$$

At this point we know

$$u_{00}(x) + v_{00}(x/\varepsilon) = 0, \tag{4.141}$$

$$u_{10}(x) + v_{10}(x/\varepsilon) = 0, \tag{4.142}$$

$$u_{11}(x) + v_{11}(x/\varepsilon) = \frac{1}{2}d_{00}x, \tag{4.143}$$

$$u_{20}(x) + v_{20}(x/\varepsilon) = d_{00}[Q(x/\varepsilon) - Q(0)] + D(x). \tag{4.144}$$

We also know $v_{21}(X) = v_{22}(X) = 0$. To determine $u_{21}(x)$ and $u_{22}(x)$, we need to calculate $Z_{31}(X)$ and $Z_{32}(X)$. For this we have

$$\widetilde{W}_3'' + 2X\widetilde{W}_3' = d_{01} + d_{10}X - c_{00}\widetilde{W}_2(X) - (b_{00} + a_1X^2)\widetilde{W}_2'(X), \tag{4.145}$$

where $b_{00} = b(0,0)$, $c_{00} = c(0,0)$, $a_1 = a'(0)$, $d_{ij} = d^{[i,j]}(0,0)$ and $\widetilde{W}_2(X)$ is given by (4.133). If we denote the right side of (4.145) by $f(X)$, then

$$\widetilde{W}_3'(X) = e^{-X^2} \int_0^X e^{T^2} f(T)\, dT \tag{4.146}$$

and

$$f(X) = \mu + \lambda X - \frac{1}{2}c_{00}d_{00}\ln(1 + X) + \phi(X), \tag{4.147}$$

where

$$\mu = d_{01} - c_{00}d_{00}Q(0), \qquad \lambda = \frac{1}{2}(2d_{10} - d_{00}a_1)X \tag{4.148}$$

and $\phi(X) \in C^\infty([0, \infty])$. Indeed,

$$\phi(X) = c_{00}d_{00}Q(X) - b_{00}d_{00}P(X) - \frac{1}{2}a_1d_{00}[2X^2P(X) - X], \tag{4.149}$$

and, in particular, $\phi(\infty) = 0$. Consequently,

$$e^{-X^2} \int_0^X e^{T^2} \phi(T) \, dT = O(X^{-2}). \tag{4.150}$$

Also

$$e^{-X^2} \int_0^X e^{T^2} \ln(1+T) \, dT = P(X) \ln(1+X) + \Pi(X), \tag{4.151}$$

where

$$\Pi(X) = e^{-X^2} \int_0^X e^{T^2} P(T) \frac{dT}{1+T}, \tag{4.152}$$

so $\Pi(X) = O(X^{-3})$ and $\Pi(X) \in C^\infty([0,\infty])$. Therefore

$$\widetilde{W}_3'(X) = \mu P(X) + \frac{1}{2}\lambda - \frac{1}{2}c_{00}d_{00}P(X)\ln(1+X) + \psi(X), \tag{4.153}$$

where $\psi(X) \in C^\infty([0,\infty])$ and $\psi(X) = O(X^{-2})$.

We already know

$$\int_0^X P(T) \, dT = Q(0) - Q(X) + \frac{1}{2}\ln(1+X). \tag{4.154}$$

In addition, it is readily determined that

$$\int_0^X P(T) \ln(1+T) \, dT = R(X) - Q(X)\ln(1+X) + \frac{1}{4}\ln^2(1+X), \tag{4.155}$$

where $R(X) \in C^\infty([0,\infty])$. Therefore (4.153) implies

$$Z_{31}(X) = \frac{1}{2}\mu X + \frac{1}{2}c_{00}d_{00}XQ(X), \qquad Z_{32}(X) = -\frac{1}{8}c_{00}d_{00}X \tag{4.156}$$

and thus from (4.58) and (4.59), we now know

$$u'_{21}(0) = \frac{1}{2}\mu, \qquad u'_{22}(0) = -\frac{1}{8}c_{00}d_{00}. \tag{4.157}$$

Using these two results in conjunction with (4.125) and (4.126), we readily find

$$u_{21}(x) + v_{21}(x/\varepsilon) = \frac{1}{2}\mu x - \frac{1}{2}d_{00}xC(x), \tag{4.158}$$

and

$$u_{22}(x) + v_{22}(x) = -\frac{1}{8}c_{00}d_{00}x, \tag{4.159}$$

where

$$C(x) = \int_0^x \left[\frac{c_0(t)}{a(t)} - \frac{c_{00}}{2} \right] \frac{dt}{t}. \tag{4.160}$$

4.5 Problem G, Part 3

When we substitute the results of the last section, beginning with (4.141), into (4.116), and also note that $\ln(1 + 1/\varepsilon) = \ln(1/\varepsilon) + O(\varepsilon)$, we get, for the solution of (4.89) with $w(0, \varepsilon) = w'(0, \varepsilon) = 0$, the uniformly valid expansion

$$w(x, \varepsilon) = \ln(1/\varepsilon)w_{11}(x, \varepsilon) + w_{20}(x, \varepsilon)$$

$$+ \varepsilon \ln^2(1/\varepsilon)w_{22}(x, \varepsilon) + \varepsilon \ln(1/\varepsilon)w_{21}(x, \varepsilon) + O(\varepsilon) \tag{4.161}$$

for $0 \le x \le 1$, where

$$w_{11}(x, \varepsilon) = \frac{1}{2}d_{00}\frac{\ln(1 + x/\varepsilon)}{\ln(1 + 1/\varepsilon)}, \tag{4.162}$$

$$w_{20}(x, \varepsilon) = D(x) + d_{00}Q(x/\varepsilon), \tag{4.163}$$

$$w_{22}(x, \varepsilon) = -\frac{1}{8}c_{00}d_{00}\frac{\ln^2(1 + x/\varepsilon)}{\ln^2(1 + 1/\varepsilon)}, \tag{4.164}$$

$$w_{21}(x, \varepsilon) = \frac{1}{2}[d_{01} - c_{00}d_{00}Q(\infty) - d_{00}C(x)]\frac{\ln(1 + x/\varepsilon)}{\ln(1 + 1/\varepsilon)} \tag{4.165}$$

The function $\tilde{z}(x, \varepsilon) = \varepsilon^2 z(x, \varepsilon)$, where $z(x, \varepsilon)$ is the desired solution to (4.90), has an expansion of the same form as the expansion (4.116) for $\tilde{w}(x, \varepsilon)$. This follows from our remarks at the end of Section 4.3. When we calculate the terms of this expansion, the result corresponding to (4.161) is

$$z(x, \varepsilon) = z_{20}(x, \varepsilon) + \varepsilon \ln(1/\varepsilon)z_{21}(x, \varepsilon) + O(\varepsilon) \tag{4.166}$$

uniformly as $\varepsilon \to 0^+$ for $0 \le x \le 1$, where

$$z_{20}(x, \varepsilon) = \text{erf}(x/\varepsilon), \tag{4.167}$$

$$z_{21}(x, \varepsilon) = -\frac{1}{2}c_{00}\frac{\ln(1 + x/\varepsilon)}{\ln(1 + 1/\varepsilon)}. \tag{4.168}$$

Returning now to (4.91), to determine $\kappa(\varepsilon)$, we have $w(1, \varepsilon) + \kappa(\varepsilon)z(1, \varepsilon) = \beta(\varepsilon)$ and a straightforward calculation, utilizing $1/z(1, \varepsilon) = 1 + \frac{1}{2}c_{00}\varepsilon \ln(1/\varepsilon) + O(\varepsilon)$, reveals

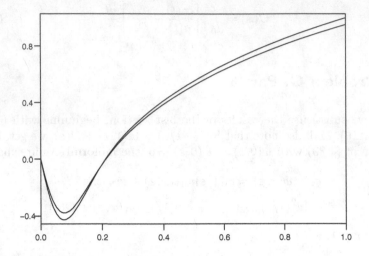

Fig. 4.1 Numerical solution to (4.88) and an asymptotic approximation to the solution when $a(x,0) = 2+x$, $b(x,\varepsilon = 0$, $c(x,\varepsilon) = \cos(x)$, $d(x,\varepsilon) = 1+x+\varepsilon$, $y(0,\varepsilon) = 0$, $y(1,\varepsilon) = 1$ and $\varepsilon = 0.1$

$$\kappa(\varepsilon) = -\frac{1}{2}d_{00}\ln(1/\varepsilon) + [\beta_0 - d_{00}Q(\infty) - D(1)]$$

$$-\frac{1}{8}c_{00}d_{00}\varepsilon \ln^2(1/\varepsilon) + O(\varepsilon\ln(1/\varepsilon)), \tag{4.169}$$

where $\beta_0 = \beta(0)$. When we substitute this, along with the above expansions for $w(x,\varepsilon)$ and $z(x,\varepsilon)$ into (4.91), we get, finally, for the solution to Problem G,

$$y(x,\varepsilon) = \ln(1/\varepsilon)y_{11}(x,\varepsilon) + y_{20}(x,\varepsilon) + \varepsilon\ln^2(1/\varepsilon)y_{22}(x,\varepsilon)$$

$$+ O(\varepsilon\ln(1/\varepsilon)), \tag{4.170}$$

where

$$y_{11}(x,\varepsilon) = \frac{1}{2}d_{00}\left[\frac{\ln(1+x/\varepsilon)}{\ln(1+1/\varepsilon)} - \text{erf}(x/\varepsilon)\right], \tag{4.171}$$

$$y_{20}(x,\varepsilon) = [\beta_0 - d_{00}Q(\infty) - D(1)]\text{erf}(x/\varepsilon) + d_{00}Q(x/\varepsilon) + D(x), \tag{4.172}$$

$$y_{22}(x,\varepsilon) = \frac{1}{8}c_{00}d_{00}\left[2\frac{\ln(1+x/\varepsilon)}{\ln(1+1/\varepsilon)} - \frac{\ln^2(1+x/\varepsilon)}{\ln^2(1+1/\varepsilon)} - \text{erf}(x/\varepsilon)\right], \tag{4.173}$$

and this holds uniformly for $0 \le x \le 1$ as $\varepsilon \to 0^+$.

A comparison of the numerical solution of (4.88) and the asymptotic result (4.170)-(4.173), assuming $a(x) = 2+x$, $b(x,\varepsilon) = 0$, $c(x,\varepsilon) = \cos(x)$, $d(x,\varepsilon) = 1+x+\varepsilon$ and $\beta(\varepsilon) = 1$, in which case $D(x) = \frac{1}{2}\ln(1+x/2)$, is shown for $\varepsilon = 0.1$

in Figure 4.1. Note also that if instead of the boundary condition $y(1, \varepsilon) = \beta(\varepsilon)$ for (4.88), we impose the initial condition $y'(0, \varepsilon) = (2/\sqrt{\pi})\beta(\varepsilon)\varepsilon^{-1}$, in addition to $y(0, \varepsilon) = 0$, then

$$y(x, \varepsilon) = \ln(1/\varepsilon)w_{11}(x, \varepsilon) + [w_{20}(x, \varepsilon) + \beta_0 z_{20}(x, \varepsilon)] + \varepsilon \ln^2(1/\varepsilon)w_{22}(x, \varepsilon)$$

$$+ \varepsilon \ln(1/\varepsilon)[w_{21}(x, \varepsilon) + \beta_0 z_{21}(x, \varepsilon)] + O(\varepsilon). \qquad (4.174)$$

4.6 Exercises

4.1. Assume $a(x, \varepsilon) \in C^\infty([0, 1] \times [0, \varepsilon_o])$ for some $\varepsilon_o > 0$, let $m \geq 0, n \geq 1$ be integers and, for $\varepsilon > 0$, let

$$y(x, \varepsilon) = \int_0^x \frac{t^n a(t, \varepsilon)}{t + \varepsilon} \ln^m \frac{1 + t/\varepsilon}{1 + x/\varepsilon} dt. \qquad (4.175)$$

The problem is to show there exists $a_k(x, \varepsilon) \in C^\infty([0, 1] \times [0, \varepsilon_o])$ for $0 \leq k \leq n - 1$ and $c_k(\varepsilon) \in C^\infty([0, \varepsilon_o])$ for $1 \leq k \leq m + 1$ such that

$$y(x, \varepsilon) = \sum_{k=0}^{n-1} \varepsilon^k x^{n-k} a_k(x, \varepsilon) + \varepsilon^n \sum_{k=1}^{m+1} c_k(\varepsilon) \ln^k(1 + x/\varepsilon) \qquad (4.176)$$

for all $(x, \varepsilon) \in [0, 1] \times (0, \varepsilon_o]$. Do this first for $m = 0$ and then, by induction, for arbitrary $m \geq 0$. Now use this to show that (4.21) follows from (4.20), where $n \geq 0, m \geq 1$.

4.2. Use the Maple routine of Section 1.5 to confirm the expansion results stated at the end of Section 4.1 for $y(x, \varepsilon)$ given by (4.27).

4.3. In Problem F, let $a(x, \varepsilon) = \cos(x)$, $b(x, \varepsilon) = 1 + \sin(x)$. Using *Prob* F, or a simplified version, show that in (4.47), by computing sufficiently many terms of just the inner expansion of $y(x, \varepsilon)$, we get both

$$v_{00}(X) = \frac{-1}{1 + X}, \qquad v_{10}(X) = \frac{1}{2(1 + X)}, \qquad v_{11}(X) = 0, \qquad (4.177)$$

$$v_{20}(X) = \frac{-4}{12(1 + X)}, \qquad v_{21}(X) = \frac{1}{2(1 + X)}, \qquad v_{22}(X) = 0, \qquad (4.178)$$

and

$$u_{00}(x) = 1 + \frac{1}{2}x + \frac{1}{6}x^2 + \frac{1}{48}x^3 + \frac{1}{120}x^4 - \frac{1}{2880}x^5 + O(x^6), \qquad (4.179)$$

$$u_{10}(x) = -\frac{1}{2} - \frac{5}{12}x - \frac{5}{48}x^2 - \frac{1}{20}x^3 - \frac{17}{2880}x^4 - \frac{79}{60480}x^5 + O(x^6), \qquad (4.180)$$

$$u_{20}(x) = \frac{11}{12} + \frac{11}{48}x + \frac{17}{90}x^2 + \frac{169}{5760}x^3 + \frac{388}{302400}x^4 + \frac{97}{120690}x^5 + O(x^6), \quad (4.181)$$

$$u_{21}(x) = -\frac{1}{2} - \frac{1}{8}x^2 - \frac{1}{96}x^4 + O(x^6), \quad (4.182)$$

and evidently $u_{11}(x) = u_{22}(x) = 0$. At least, $u_{11}(x) = O(x^6)$, $u_{22}(x) = O(x^6)$.

4.4. Work out the details for the proof of Proposition 2.

4.5. Show that the solution to (4.90) satisfying $z(0, \varepsilon) = 0$, $z'(0, \varepsilon) = (2/\sqrt{\pi})\varepsilon^{-1}$ has an infinite series solution of the same form as the one for $w(x, \varepsilon)$, and from there show that $\widetilde{z}(x, \varepsilon) = \varepsilon^2 z(x, \varepsilon)$ has a uniformly valid asymptotic expansion of the same form as $\widetilde{w}(x, \varepsilon)$. Compute the first several terms of this expansion and confirm the results (4.166)-(4.168).

Chapter 5
Oscillation Problems

5.1 Problem H

To begin this chapter, suppose

$$\varepsilon y' + a(x, \varepsilon)y = b(x, \varepsilon)\cos(x/\varepsilon), \tag{5.1}$$

where $a(x, \varepsilon)$, $b(x, \varepsilon) \in C^\infty([0, 1] \times [0, \varepsilon_o])$ for some $\varepsilon_o > 0$. Assume also that $a(x, \varepsilon) > 0$ and $y(0, \varepsilon) = 0$. We know from Problem A that

$$y(x, \varepsilon) = \varepsilon^{-1} \int_0^x f(x, t, t/\varepsilon, \varepsilon)\cos((x - t)/\varepsilon)\, dt, \tag{5.2}$$

where

$$f(x, t, T, \varepsilon) = b(x - t, \varepsilon)e^{-Tu(x,t,\varepsilon)}, \tag{5.3}$$

with

$$u(x, t, \varepsilon) = t^{-1}\int_{x-t}^x a(s, \varepsilon)\, ds, \tag{5.4}$$

and we could expand $f(x, t, t/\varepsilon, \varepsilon)$ in accordance with Corollary 2, but this would not be immediately useful. Instead, assuming $b(x, \varepsilon)$ is real valued, like $a(x, \varepsilon) > 0$, we note from (5.2) that $y(x, \varepsilon) = \text{Re}[w(x, \varepsilon)]$, where $w(x, \varepsilon) = z(x, \varepsilon)e^{ix/\varepsilon}$ and, if we let

$$\phi(x, t, T, \varepsilon) = b(x - t, \varepsilon)e^{-Tv(x,t,\varepsilon)}, \tag{5.5}$$

where

$$v(x, t, \varepsilon) = u(x, t, \varepsilon) + i, \tag{5.6}$$

then

$$z(x, \varepsilon) = \varepsilon^{-1} \int_0^x \phi(x, t, t/\varepsilon, \varepsilon)\, dt. \tag{5.7}$$

Now, since $\mathrm{Re}[v(x, t, \varepsilon)] = u(x, t, \varepsilon) > 0$ for all $(x, t, \varepsilon) \in [0, 1] \times [0, x] \times [0, \varepsilon_o]$, we have $\phi(x, t, T, \varepsilon) \in C^\infty([0, 1] \times [0, x] \times [0, \infty] \times [0, \varepsilon_o])$, and therefore, by Corollary 2, as in Problem A, there exists $u_n(x) \in C^\infty([0, 1])$ and $v_n(X) \in C^\infty([0, \infty])$ with $v_n(\infty) = 0$ such that for any $N \geq 0$,

$$z(x, \varepsilon) = \sum_{n=0}^{N-1} \varepsilon^n [u_n(x) + v_n(x/\varepsilon)] + O(\varepsilon^N) \tag{5.8}$$

uniformly as $\varepsilon \to 0^+$ for $0 \leq x \leq 1$. Furthermore, we can get the terms of this expansion, again as in Problem A, from the differential equation for $z(x, \varepsilon)$, which is

$$\varepsilon z' + [a(x, \varepsilon) + i]z = b(x, \varepsilon). \tag{5.9}$$

Thus we immediately see

$$O_1 z(x, \varepsilon) = \frac{b(x, 0)}{a(x, 0) + i} \tag{5.10}$$

and upon introducing $X = x/\varepsilon$ in (5.9), we readily find

$$I_1 z(x, \varepsilon) = \frac{b(0, 0)}{a(0, 0) + i} \left[1 - e^{-[a(0,0)+i]x/\varepsilon}\right]. \tag{5.11}$$

Consequently,

$$u_0(x) = \frac{b(x, 0)}{a(x, 0) + i}, \qquad v_0(X) = -\frac{b(0, 0)}{a(0, 0) + i} e^{-[a(0,0)+i]X}, \tag{5.12}$$

and from here it follows that

$$y(x, \varepsilon) = p_0(x) \cos(x/\varepsilon) + q_0(x) \sin(x/\varepsilon) + r_0(x/\varepsilon) + O(\varepsilon), \tag{5.13}$$

where

$$p_0(x) = \frac{a(x, 0)b(x, 0)}{1 + [a(x, 0)]^2}, \qquad q_0(x) = \frac{b(x, 0)}{1 + [a(x, 0)]^2}, \tag{5.14}$$

and

$$r_0(X) = -p_0(0)e^{-a(0,0)X}. \tag{5.15}$$

In general, if $y(x, \varepsilon)$ is the solution to (5.1), where $a(x, \varepsilon)$, $b(x, \varepsilon)$ are real valued, $a(x, \varepsilon) > 0$ and $y(0, \varepsilon) = 0$, then

$$y(x, \varepsilon) = p(x, \varepsilon) \cos(x/\varepsilon) + q(x, \varepsilon) \sin(x/\varepsilon) + r(x, \varepsilon), \tag{5.16}$$

where

$$p(x, \varepsilon) = \sum_{n=0}^{N-1} \varepsilon^n p_n(x) + O(\varepsilon^N), \qquad q(x, \varepsilon) = \sum_{n=0}^{N-1} \varepsilon^n q_n(x) + O(\varepsilon^N), \tag{5.17}$$

and

$$r(x, \varepsilon) = \sum_{n=0}^{N-1} \varepsilon^n r_n(x/\varepsilon) + O(\varepsilon^N) \tag{5.18}$$

all uniformly as $\varepsilon \to 0^+$ for $0 \leq x \leq 1$, where $p_n(x)$, $q_n(x) \in C^\infty([0, 1])$, and $r_n(X) \in C^\infty([0, \infty])$. Furthermore, in specific cases, the terms of these expansions can be determined with the help of *ProbA*, as in Exercise 5.1.

The asymptotic expansion of an integral such as (5.7), where $\phi(x, t, T, \varepsilon)$ is given by (5.5), is rather different if $\mathrm{Re}[v(x, t, \varepsilon)] = 0$. We will need the proposition below in the next section.

Proposition 3. *Assume $g(x)$, $h(x) \in C^\infty([0, 1])$ are real valued, $h(0) = 0$, and*

$$F(x, \varepsilon) = \varepsilon^{-1} \int_0^x e^{-ih(t)/\varepsilon} g(t) \, dt. \tag{5.19}$$

(a). If $h'(x) \neq 0$ for $0 \leq x \leq 1$, then there exists $c(\varepsilon)$, $u_n(x) \in C^\infty([0, 1])$ such that

$$F(x, \varepsilon) = c(\varepsilon) + \phi(x, \varepsilon) e^{-ih(x)/\varepsilon}, \tag{5.20}$$

where, for any $N \geq 0$,

$$\phi(x, \varepsilon) = \sum_{n=0}^{N-1} \varepsilon^n u_n(x) + O(\varepsilon^N) \tag{5.21}$$

uniformly as $\varepsilon \to 0^+$ for $0 \leq x \leq 1$.

(b). If $h'(x) > 0$ for $0 < x \leq 1$, $h'(0) = 0$ and $h''(0) > 0$, then there exists $c(\hat{\varepsilon})$, $u_n(x) \in C^\infty([0, 1])$ and $v_n(X) \in C^\infty([0, \infty])$ such that

$$\hat{\varepsilon} F(x, \varepsilon) = c(\hat{\varepsilon}) + \phi(x, \hat{\varepsilon}) e^{-ih(x)/\varepsilon} \tag{5.22}$$

where $\hat{\varepsilon} = \varepsilon^{1/2}$ and, for any $N \geq 0$,

$$\phi(x, \hat{\varepsilon}) = \sum_{n=0}^{N-1} \hat{\varepsilon}^n [u_n(x) + v_n(x/\hat{\varepsilon})] + O(\hat{\varepsilon}^N) \tag{5.23}$$

uniformly as $\hat{\varepsilon} \to 0^+$ for $0 \leq x \leq 1$.

Proof. If $h'(x) \neq 0$ for $0 \leq x \leq 1$, then $u_0(x) = -ig(x)/h'(x)$ is in $C^\infty([0, 1])$ and an integration by parts yields

$$F(x, \varepsilon) = u_0(0) - u_0(x) e^{-ih(x)/\varepsilon} + \varepsilon F_1(x, \varepsilon), \tag{5.24}$$

where

$$F_1(x, \varepsilon) = \varepsilon^{-1} \int_0^x e^{-ih(x)/\varepsilon} u_0'(t) \, dt. \tag{5.25}$$

But this means

$$F_1(x, \varepsilon) = u_1(0) - u_1(x)e^{-ih(x)/\varepsilon} + \varepsilon F_2(x, \varepsilon), \tag{5.26}$$

where $u_1(x) = -iu_0'(x)/h'(x)$ and

$$F_2(x, \varepsilon) = \varepsilon^{-1} \int_0^x e^{-ih(t)/\varepsilon} u_1'(t)\, dt. \tag{5.27}$$

Obviously we can repeat this process indefinitely, and this proves Part (a) of the proposition. Indeed, we have $u_n(x) = -iu_{n-1}'(x)/h'(x)$ for $n \geq 1$ and we can choose $c(\varepsilon)$ to be any function in $C^\infty([0,1])$ such that $c^{[n]}(0) = u_n(0)$.

To prove Part (b), let $s = [h(x)]^{1/2}$, let $x = \theta(s)$ denote the inverse of this transformation, and let $G(s) = g(\theta(s))\theta'(s)$. Then

$$\hat{\varepsilon}F(x, \varepsilon) = \hat{\varepsilon}^{-1} \int_0^{[h(x)]^{1/2}} e^{-i(s/\hat{\varepsilon})^2} [G(0) + 2isG_1(s)]\, ds, \tag{5.28}$$

where $G_1(s) = (2is)^{-1}[G(s) - G(0)]$, and an integration by parts reveals

$$\hat{\varepsilon}F(x, \varepsilon) = H(0, \varepsilon) - H([h(x)]^{1/2}, \varepsilon) + \varepsilon F_1(x, \varepsilon), \tag{5.29}$$

where

$$H(x, \varepsilon) = e^{-ix^2/\varepsilon}[\Psi(x/\hat{\varepsilon})G(0) + \hat{\varepsilon}G_1(x)], \tag{5.30}$$

$$\Psi(x) = e^{ix^2} \int_x^\infty e^{-is^2}\, ds, \tag{5.31}$$

and

$$F_1(x, \varepsilon) = \hat{\varepsilon}^{-1} \int_0^{[h(x)]^{1/2}} e^{-i(s/\hat{\varepsilon})^2} [G_1'(0) + 2isG_2(s)]\, ds. \tag{5.32}$$

where $G_2(s) = (2is)^{-1}[G_1'(s) - G_1'(0)]$. So again we have a process we can iterate indefinitely. Furthermore, $\Psi(x) \in C^\infty([0, \infty])$, as we see in Exercise 5.2, and therefore we can expand $\Psi([h(x)]^{1/2}/\hat{\varepsilon}) = \Psi((x/\hat{\varepsilon})[h(x)/x^2]^{1/2})$ according to Theorem 1. This leads directly to (5.22)-(5.23).

5.2 Problem I, Part 1

For our final problem, we seek an asymptotic expansion for the solution to

$$\varepsilon^2 y'' + a(x, \varepsilon)y = \varepsilon^{1/2} b(x, \varepsilon)\cos(x/\varepsilon) \tag{5.33}$$

satisfying the initial conditions $y(0, \varepsilon) = \alpha(\varepsilon)$, $y'(0, \varepsilon) = \varepsilon^{-1}\beta(\varepsilon)$. The functions $a(x, \varepsilon)$, $b(x, \varepsilon) \in C^\infty([0, 1] \times [0, \varepsilon_o])$ and $\alpha(\varepsilon)$, $\beta(\varepsilon) \in C^\infty([0, \varepsilon_o])$ for some $\varepsilon_o > 0$ are assumed to be real, and we assume $a(x, \varepsilon) > 0$. We further assume, for some $x_o \in (0, 1)$, that $a_{00}(x) < 1$ for $0 \le x < x_o$ and $a_{00}(x) > 1$ for $x_o < x \le 1$, where, in general, $a_{ij}(x) = a^{[i,j]}(x, 0)$, $b_{ij}(x) = b^{[i,j]}(x, 0)$. Finally, we assume $a_{10}(x_o) \ne 0$.

If (instead) $a(x, \varepsilon) = (1 + \varepsilon c)^2$, where c is a constant, and $b(x, \varepsilon) = \varepsilon^{1/2}$, then

$$y(x, \varepsilon) = \alpha(\varepsilon) \cos[(1 + \varepsilon c)(x/\varepsilon)] + \frac{\beta(\varepsilon)}{1 + \varepsilon c} \sin[(1 + \varepsilon c)(x/\varepsilon)] + y_p(x, \varepsilon), \quad (5.34)$$

where, in the particular solution

$$y_p(x, \varepsilon) = M(x, \varepsilon) \sin(x/\varepsilon) + N(x, \varepsilon) \cos(x/\varepsilon), \quad (5.35)$$

assuming $|c| << 1/\varepsilon$, the amplitudes

$$M(x, \varepsilon) = \frac{\sin(cx)}{c(2 + \varepsilon c)}, \qquad N(x, \varepsilon) = \frac{1 - \cos(cx)}{c(2 + \varepsilon c)} \quad (5.36)$$

oscillate slowly compared to $\sin(x/\varepsilon)$ and $\cos(x/\varepsilon)$, unless $c = 0$, in which case $M(x, \varepsilon) = x/2$, $N(x, \varepsilon) = 0$. The point of Problem I is see what happens if $a(x, \varepsilon)$ comes close to 1 but only momentarily equals 1 as x increases from 0 to 1.

As in Problem II, it is convenient to let $w(x, \varepsilon)$ be the solution to (5.33) with $e^{ix/\varepsilon}$ in place of $\cos(x/\varepsilon)$, so that $y(x, \varepsilon) = \text{Re}[w(x, \varepsilon)]$. In analogy with Problem D, as in [11], the homogeneous equation for $w(x, \varepsilon)$ has a pair of linearly independent solutions,

$$w_h^{(1)}(x, \varepsilon) = e^{is(x)/\varepsilon} u(x, \varepsilon), \qquad w_h^{(2)}(x, \varepsilon) = \bar{w}_h^{(1)}(x, \varepsilon), \quad (5.37)$$

where $\bar{w}_h^{(1)}(x, \varepsilon)$ is the complex conjugate of $w_h^{(1)}(x, \varepsilon)$,

$$s(x) = \int_0^x [a_{00}(t)]^{1/2} \, dt, \quad (5.38)$$

and $u(x, \varepsilon)$ has a uniformly valid asymptotic expansion,

$$u(x, \varepsilon) = \sum_{n=0}^{N-1} \varepsilon^n u_n(x) + O(\varepsilon^N), \quad (5.39)$$

for $0 \le x \le 1$, where $u_n(x) \in C^\infty([0, 1])$. In particular, if we take $u(0, \varepsilon) = 1$, it is readily determined that

$$u_0(x) = [a_{00}(0)/a_{00}(x)]^{1/4} e^{ir(x)}, \quad (5.40)$$

where

$$r(x) = \frac{1}{2} \int_0^x [a_{00}(t)]^{-1/2} a_{01}(t) \, dt. \tag{5.41}$$

Furthermore, the Wronskian of these two solutions is $\varepsilon^{-1} \Delta(\varepsilon)$, where $\Delta(\varepsilon)$ is pure imaginary and also has an asymptotic expansion in powers of ε. In particular, $\Delta(0) = -2i[a_{00}(0)]^{1/2}$. It follows by the method of variation of parameters that the full equation for $w(x, \varepsilon)$ has a particular solution of the form

$$w_p(x, \varepsilon) = A(x, \varepsilon) w_h^{(1)}(x, \varepsilon) + B(x, \varepsilon) w_h^{(2)}(x, \varepsilon) \tag{5.42}$$

where

$$A(x, \varepsilon) = \varepsilon^{-1/2} \int_{x_o}^x e^{-i[s(t)-t]/\varepsilon} \bar{\zeta}(t, \varepsilon) \, dt, \tag{5.43}$$

$$B(x, \varepsilon) = \varepsilon^{-1/2} \int_{x_o}^x e^{i[s(t)+t]/\varepsilon} \zeta(t, \varepsilon) \, dt, \tag{5.44}$$

with $\zeta(x, \varepsilon) = b(x, \varepsilon) u(x, \varepsilon)/\Delta(\varepsilon)$.

If we substitute $t = \hat{t} + x_o$ in (5.43) and let $h(\hat{t}) = s(\hat{t} + x_o) - s(x_o) - \hat{t}$, then $h'(\hat{t}) > 0$ for $0 < \hat{t} \leq 1 - x_o$, $h(0) = h'(0) = 0$ and $h''(0) = \frac{1}{2} a'(x_o) > 0$. Therefore, by Part (b) of Proposition 3,

$$A(x, \varepsilon) = e^{-i[s(x_o) - x_o]/\varepsilon} [c(\hat{\varepsilon}) + \phi(\hat{x}, \hat{\varepsilon}) e^{-ih(\hat{x})/\varepsilon}], \tag{5.45}$$

where $\hat{\varepsilon} = \varepsilon^{1/2}$, $\hat{x} = x - x_o$, $c(\hat{\varepsilon}) \in C^\infty([0, 1])$ and for any $N \geq 0$,

$$\phi(\hat{x}, \hat{\varepsilon}) = \sum_{n=0}^{N-1} \hat{\varepsilon}^n [u_n(\hat{x}) + v_n(\hat{x}/\hat{\varepsilon})] + O(\hat{\varepsilon}^N) \tag{5.46}$$

uniformly as $\hat{\varepsilon} \to 0^+$ for $0 \leq \hat{x} \leq 1 - x_o$, that is, $x_o \leq x \leq 1$, where $u_n(\hat{x}) \in C^\infty([0, 1 - x_o])$, $v_n(\hat{X}) \in C^\infty([0, \infty])$. Furthermore, $-[s(x_o) - x_o] - h(\hat{x}) = -s(x) + x$ and therefore, by an application of Theorem 1, there exists $c_A(\hat{\varepsilon})$ and $\phi_A(\hat{x}, \hat{\varepsilon})$ such that

$$A(x, \varepsilon) w_h^{(1)}(x, \varepsilon) = c_A(\hat{\varepsilon}) w_h^{(1)}(x, \varepsilon) + \phi_A(\hat{x}, \hat{\varepsilon}) e^{ix/\varepsilon}, \tag{5.47}$$

where $\phi_A(\hat{x}, \hat{\varepsilon})$ has a uniformly valid expansion of the same form as $\phi(\hat{x}, \hat{\varepsilon})$. Similarly, by Part (a) of Proposition 5.1, we find

$$B(x, \varepsilon) w_h^{(2)}(x, \varepsilon) = c_B(\hat{\varepsilon}) w_h^{(2)}(x, \varepsilon) + \phi_B(\hat{x}, \hat{\varepsilon}) e^{ix/\varepsilon}, \tag{5.48}$$

where $\phi_B(\hat{x}, \hat{\varepsilon})$ also has a uniformly valid expansion like $\phi(\hat{x}, \hat{\varepsilon})$, although there are no $v_n(\hat{x}/\hat{\varepsilon})$ terms. Consequently, when we substitute (5.47) and (5.48) into (5.42) we see there is a particular solution $w^{(+)}(x, \varepsilon) = z^{(+)}(\hat{x}, \hat{\varepsilon}) e^{ix/\varepsilon}$ of the differential equation for $w(x, \varepsilon)$, in addition to $w_p(x, \varepsilon)$, such that, for certain $u_n^{(+)}(\hat{x}) \in C^\infty([0, 1 - x_o])$, $v_n^{(+)}(\hat{X}) \in C^\infty([0, \infty])$,

$$z^{(+)}(\hat{x}, \hat{\varepsilon}) = \sum_{n=0}^{N-1} \hat{\varepsilon}^n [u_n^{(+)}(\hat{x}) + v_n^{(+)}(\hat{x}/\hat{\varepsilon})] + O(\hat{\varepsilon}^N) \tag{5.49}$$

uniformly as $\varepsilon \to 0^+$ for $0 \le \hat{x} \le 1 - x_o$.

The differential equation for $z^{(+)}(\hat{x}, \hat{\varepsilon})$ is

$$\hat{\varepsilon}^4 z^{(+)\prime\prime} + 2i\hat{\varepsilon}^2 z^{(+)\prime} + [a(x_o + \hat{x}, \hat{\varepsilon}^2) - 1]z^{(+)} = \hat{\varepsilon} b(x_o + \hat{x}, \hat{\varepsilon}^2) \tag{5.50}$$

and for the first term of the N-term outer expansion,

$$O_N z^{(+)}(\hat{x}, \hat{\varepsilon}) = \sum_{n=0}^{N-1} \hat{\varepsilon}^n z_n^{(+)}(\hat{x}), \tag{5.51}$$

whose existence is implied by (5.49), it is apparent that $z_0^{(+)}(\hat{x}) = 0$. Also, for the leading term of the N-term inner expansion of $z^{(+)}(\hat{x}, \hat{\varepsilon})$,

$$I_N z^{(+)}(\hat{x}, \hat{\varepsilon}) = \sum_{n=0}^{N-1} \hat{\varepsilon}^n Z_n^{(+)}(\hat{X}), \tag{5.52}$$

it is clear from the differential equation

$$\hat{\varepsilon}^2 Z^{(+)\prime\prime} + 2i\hat{\varepsilon} Z^{(+)\prime} + [a(x_o + \hat{\varepsilon}\hat{X}, \hat{\varepsilon}^2) - 1]Z^{(+)} = \hat{\varepsilon} b(x_o + \hat{\varepsilon}\hat{X}, \hat{\varepsilon}^2) \tag{5.53}$$

for $Z^{(+)}(\hat{X}, \hat{\varepsilon}) = z^{(+)}(x_o + \hat{\varepsilon}\hat{X}, \hat{\varepsilon})$ that

$$2i Z_0^{(+)\prime\prime} + a_{10}(x_o)\hat{X} Z_0^{(+)} = b_{00}(x_o). \tag{5.54}$$

Thus, in terms of the function $\Psi(x) \in C^\infty([0, \infty])$ introduced in the proof of Proposition 3, and further discussed in Exercise 5.2, in order for $O_1 I_1 z^{(+)}(\hat{x}, \hat{\varepsilon})$ to agree with the fact that $I_1 O_1 z^{(+)}(\hat{x}, \hat{\varepsilon}) = 0$, it must be that

$$Z_0^{(+)}(\hat{X}) = \frac{ib_{00}(x_o)}{2m_o} \Psi(m_o \hat{X}), \tag{5.55}$$

where $m_o = \frac{1}{2}[a_{10}(x_o)]^{1/2}$. In other words, in (5.49),

$$u_0^{(+)}(\hat{x}) = 0, \qquad v_0^{(+)}(\hat{X}) = Z_0^{(+)}(\hat{X}). \tag{5.56}$$

5.3 Problem I, Part 2

If we put $t = x_o - \tilde{t}$ in (5.43) and let $h(\tilde{t}) = s(x_o - \tilde{t}) - s(x_o) + \tilde{t}$, then by another application of Proposition 3, Part (b), we see, in addition to (5.45), that

$$A(x, \varepsilon) = e^{-i[s(x_o) - x_o]/\varepsilon}[d(\hat{\varepsilon}) + \psi(\tilde{x}, \hat{\varepsilon})e^{-ih(\tilde{x})/\varepsilon}], \tag{5.57}$$

where $\tilde{x} = x_o - x$, $d(\hat{\varepsilon}) \in C^\infty([0, 1])$ and $\psi(\tilde{x}, \hat{\varepsilon})$ has a uniformly valid asymptotic expansion for $0 \leq \tilde{x} \leq x_o$ of the same form as the one for $\phi(\hat{x}, \hat{\varepsilon})$ on $0 \leq \hat{x} \leq 1 - x_o$. A similar statement applies to $B(x, \varepsilon)$ and thus we see, in addition to the particular solution $w^{(+)}(x, \varepsilon) = z^{(+)}(\hat{x}, \hat{\varepsilon})e^{ix/\varepsilon}$ to the differential equation for $w(x, \varepsilon)$, there is another particular solution, $w^{(-)}(x, \varepsilon) = z^{(-)}(\tilde{x}, \hat{\varepsilon})e^{ix/\varepsilon}$, such that, for certain $u_n^{(-)}(\tilde{x}) \in C^\infty([0, x_o])$, $v_n^{(-)}(\tilde{X}) \in C^\infty([0, \infty])$,

$$z^{(-)}(\tilde{x}, \hat{\varepsilon}) = \sum_{n=0}^{N-1} \hat{\varepsilon}^n [u_n^{(-)}(\tilde{x}) + v_n^{(-)}(\tilde{x}/\hat{\varepsilon})] + O(\hat{\varepsilon}^N), \tag{5.58}$$

uniformly as $\varepsilon \to 0^+$ for $0 \leq \tilde{x} \leq x_o$. Furthermore, it is readily determined from the differential equation for $z^{(-)}(\tilde{x}, \hat{\varepsilon})$, in analogy with the calculations for $z^{(+)}(\hat{x}, \hat{\varepsilon})$, that

$$u_0^{(-)}(\tilde{x}) = 0, \qquad v_0^{(-)}(\tilde{X}) = -v_0^{(+)}(\tilde{X}). \tag{5.59}$$

To solve now for $y(x, \varepsilon) = \text{Re}[w(x, \varepsilon)]$, first we note that, since the difference between any two particular solutions of the differential equation for $w(x, \varepsilon)$ is a solution to the corresponding homogeneous differential equation, there exists $c_1(\varepsilon)$, $c_2(\varepsilon)$ such that

$$w^{(-)}(x, \varepsilon) = w^{(+)}(x, \varepsilon) + c_1(\varepsilon)w_h^{(1)}(x, \varepsilon) + c_2(\varepsilon)w_h^{(2)}(x, \varepsilon). \tag{5.60}$$

Furthermore, to determine $c_1(\varepsilon)$ and $c_2(\varepsilon)$ we have, in view of (5.56) and (5.59),

$$w^{(-)}(x_o, \varepsilon) - w^{(+)}(x_o, \varepsilon) = [2Z^{(-)}(0) + O(\hat{\varepsilon})]e^{ix_o/\varepsilon}, \tag{5.61}$$

and also

$$w^{(-)\prime}(x_o, \varepsilon) - w^{(+)\prime}(x_o, \varepsilon) = \varepsilon^{-1}[2iZ_0^{(-)}(0) + O(\hat{\varepsilon})]e^{ix_o/\varepsilon}. \tag{5.62}$$

In addition,

$$w_h^{(1)}(x_o, \varepsilon) = [u_0(x_o) + O(\varepsilon)]e^{is(x_o)/\varepsilon}, \tag{5.63}$$

$$w_h^{(1)\prime}(x_o, \varepsilon) = \varepsilon^{-1}[iu_0(x_o) + O(\varepsilon)]e^{is(x_o)/\varepsilon}, \tag{5.64}$$

and, of course, $w_h^{(2)}(x_o, \varepsilon) = \bar{w}_h^{(1)}(x_o, \varepsilon)$. Therefore, from (5.60) and its derivative evaluated at $x = x_o$, and the fact that $\Psi(0) = (\sqrt{\pi}/2)e^{i\pi/4}$, what we find is

$$c_1(\varepsilon) = \frac{\sqrt{\pi}b_{00}(x_o)}{2m_o[a_{00}(0)]^{1/4}}e^{i\{-3\pi/4 + [x_o - s(x_o)]/\varepsilon - r(x_o)\}} + O(\hat{\varepsilon}) \tag{5.65}$$

and $c_2(\varepsilon) = O(\hat{\varepsilon})$. Consequently,

$$\operatorname{Re}[w^{(-)}(x,\varepsilon)] = \operatorname{Re}[w^{(+)}(x,\varepsilon)] + [\rho_0(x,\varepsilon) + O(\hat{\varepsilon})]\cos[r(x) + s(x)/\varepsilon]$$

$$+ [\sigma_0(x,\varepsilon) + O(\hat{\varepsilon})]\sin[r(x) + s(x)/\varepsilon], \qquad (5.66)$$

where

$$\rho_0(x,\varepsilon) = \frac{\sqrt{\pi}b_{00}(x_o)}{2m_o[a_{00}(x)]^{1/4}}\cos\{3\pi/4 - [x_o - s(x_o)]/\varepsilon + r(x_o)\}, \qquad (5.67)$$

$$\sigma_0(x,\varepsilon) = \frac{\sqrt{\pi}b_{00}(x_o)}{2m_o[a_{00}(x)]^{1/4}}\sin\{3\pi/4 - [x_o - s(x_o)]/\varepsilon + r(x_o)\}. \qquad (5.68)$$

Similarly, the difference between the particular solution $w^{(-)}(x,\varepsilon)$ and the one that satisfies the initial conditions $w(0,\varepsilon) = \alpha(\varepsilon)$, $w'(0,\varepsilon) = \varepsilon^{-1}\beta(\varepsilon)$ also is a solution to the corresponding homogeneous differential equation. Thus we find, for the solution to Problem I,

$$y(x,\varepsilon) = \operatorname{Re}[w^{(-)}(x,\varepsilon)] + [\lambda_0(x) + O(\hat{\varepsilon})]\cos[r(x) + s(x)/\varepsilon]$$

$$+ [\mu_0(x) + O(\hat{\varepsilon})]\sin[r(x) + s(x)/\varepsilon], \qquad (5.69)$$

where, with $\alpha_0 = \alpha(0)$ and $\beta_0 = \beta(0)$,

$$\lambda_0(x) = \alpha_0[a_{00}(0)/a_{00}(x)]^{1/4}, \qquad \mu_0(x) = \beta_0[a_{00}(0)a_{00}(x)]^{-1/4}. \qquad (5.70)$$

To deal with $\operatorname{Re}[w^{(-)}(x,\varepsilon)]$ and $\operatorname{Re}[w^{(+)}(x,\varepsilon)]$, note that $\Psi(x) = p(x) + iq(x)$, where

$$p(x) = \int_x^\infty (\cos x^2 \cos s^2 + \sin x^2 \sin s^2)\,ds, \qquad (5.71)$$

$$q(x) = \int_x^\infty (\sin x^2 \cos s^2 - \cos x^2 \sin s^2)\,ds. \qquad (5.72)$$

It follows that

$$\operatorname{Re}[w^{(-)}(x,\varepsilon)] = \frac{b_{00}(x_o)}{2m_o}\Big[[q(m_o(x_o - x)/\hat{\varepsilon}) + O(\hat{\varepsilon})]\cos(x/\varepsilon)$$

$$+ [p(m_o(x_o - x)/\hat{\varepsilon}) + O(\hat{\varepsilon})]\sin(x/\varepsilon)\Big] \qquad (5.73)$$

for $0 \le x \le x_o$, and

$$\operatorname{Re}[w^{(+)}(x,\varepsilon)] = -\frac{b_{00}(x_o)}{2m_o}\Big[[q(m_o(x - x_o)/\hat{\varepsilon}) + O(\hat{\varepsilon})]\cos(x/\varepsilon)$$

$$+ [p(m_o(x - x_o)/\hat{\varepsilon}) + O(\hat{\varepsilon})]\sin(x/\varepsilon)\Big]. \qquad (5.74)$$

for $x_o \le x \le 1$. Finally then, when we combine everything, we have

$$y(x,\varepsilon) = A(x,\varepsilon)\cos[r(x) + s(x)/\varepsilon] + B(x,\varepsilon)\sin[r(x) + s(x)/\varepsilon]$$

$$+ C(x,\varepsilon)\cos(x/\varepsilon) + D(x,\varepsilon)\sin(x/\varepsilon), \qquad (5.75)$$

where $A(x,\varepsilon)$ has a uniformly valid expansion

$$A(x,\varepsilon) = \sum_{n=0}^{N-1} \hat{\varepsilon}^n A_n(x,\varepsilon) + O(\hat{\varepsilon}^N) \qquad (5.76)$$

stemming from (5.49), (5.58) and (5.39), in which each $A_n(x,\varepsilon) = O(1)$ for $0 \leq x \leq 1$, and there are corresponding expansions for $B(x,\varepsilon)$, $C(x,\varepsilon)$ and $D(x,\varepsilon)$. In particular, we have determined

$$A_0(x,\varepsilon) = \lambda_0(x) + H(x - x_o)\sigma_0(x,\varepsilon), \qquad (5.77)$$

$$B_0(x,\varepsilon) = \mu_0(x) + H(x - x_o)\rho_0(x,\varepsilon), \qquad (5.78)$$

and

$$C_0(x,\varepsilon) = \frac{b_{00}(x_o)}{2m_o}q(m_o|x - x_o|/\hat{\varepsilon})\mathrm{sgn}(x - x_o) \qquad (5.79)$$

$$D_0(x,\varepsilon) = \frac{b_{00}(x_o)}{2m_o}p(m_o|x - x_o|/\hat{\varepsilon})\mathrm{sgn}(x - x_o), \qquad (5.80)$$

where $H(x) = 0$ for $x < 0$, $H(x) = 1$ for $x > 0$, and $\mathrm{sgn}(x) = H(x) - H(-x)$. In Figure 5.1, a graph of the uniformly valid $O(1)$ approximation

$$y_0(x,\varepsilon) = A_0(x,\varepsilon)\cos[r(x) + s(x)/\varepsilon)] + B_0(x,\varepsilon)\sin[r(x) + s(x)/\varepsilon]$$

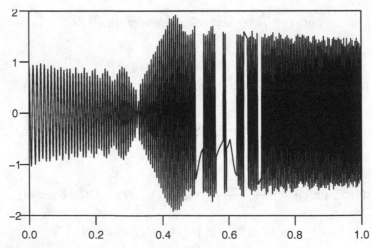

Fig. 5.1 Uniformly valid asymptotic approximation of the solution to (5.33) when $a(x,\varepsilon) = 1/4 + 2x + \varepsilon$, $b(x,\varepsilon) = 1$, $y(0,\varepsilon) = 1)$, $y'(0,\varepsilon) = 0$ and $\varepsilon = 0.001$.

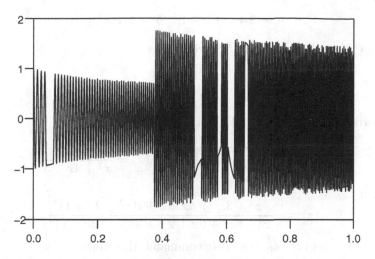

Fig. 5.2 Outer approximation of the approximate solution to (5.33) shown in Figure 5.1.

$$+ C_0(x, \varepsilon) \cos(x/\varepsilon) + D_0(x, \varepsilon) \sin(x/\varepsilon) \tag{5.81}$$

is shown for the case

$$a(x, \varepsilon) = \frac{1}{4} + 2x + \varepsilon, \quad b(x, \varepsilon) = 1, \quad \alpha(\varepsilon) = 1, \quad \beta(\varepsilon) = 0, \quad \varepsilon = 0.001. \tag{5.82}$$

Although Maple's plotting routine obviously is being pushed to the limit here, it is easy to check that the result in this case is virtually indistinguishable from Maple's numerical solution of (5.33) . Also, on any subinterval of $[0, 1]$ excluding the point $x = x_o$, the inner expansion terms $C_0(x, \varepsilon)$ and $D_0(x, \varepsilon)$ are uniformly $O(\hat{\varepsilon})$ and therefore $y(x, \varepsilon) = y_{00}(x, \varepsilon) + O(\hat{\varepsilon})$, where

$$y_{00}(x, \varepsilon) = A_0(x, \varepsilon) \cos[r(x) + s(x)/\varepsilon)] + B_0(x, \varepsilon) \sin[r(x) + s(x)/\varepsilon]. \tag{5.83}$$

A graph of the (outer) approximation $y_{00}(x, \varepsilon)$, again assuming (5.82), is shown in Figure 5.2. In general, the amplitude jump in $y_{00}(x, \varepsilon)$ at $x = x_o$ is from $[\lambda_0(x_o)^2 + \mu_0(x_o)^2]^{1/2}$ to the square root of $[\lambda_0(x_o) + \rho_0(x_o)]^2 + [\mu_0(x_o) + \sigma_0(x_o, \varepsilon)]^2$, and in the case of (5.82), this jump is from $\sqrt{2}/2$ to $[(1 + \pi)/2 + \sqrt{\pi} \cos(3\pi/4 - 997/12)]^{1/2}$.

5.4 Exercises

5.1. Specific solutions to Problem H quickly become unwieldy. For example, if $a(x, \varepsilon) = 2 + x + \varepsilon$ and $b(x, \varepsilon) = 1 + x^2$, using *ProbA* with   •

$$a := 2 + x + \varepsilon + \text{I}; \quad b := 1 + x^2 \tag{5.84}$$

to solve (5.9) for the terms of (5.8) yields

$$u_0(x) = \frac{1 + x^2}{x + 2 + i} \tag{5.85}$$

and therefore, in agreement with (5.14),

$$p_0(x) = \frac{x^3 + 2x^2 + x + 2}{x^2 + 4x + 5}, \qquad q_0(x) = \frac{x^2 + 1}{x^2 + 4x + 5}. \tag{5.86}$$

But then

$$u_1(x) = -\frac{x^3 + 3x^2 + 5x + 1 + i(x^2 + 2x + 1)}{x^3 + 6x^2 + 9x + 2 + i(3x^2 + 12x + 11)} \tag{5.87}$$

and, with the help of Maple's *evalc* command, this implies

$$p_1(x) = -\frac{x^6 + 9x^5 + 35x^4 + 78x^3 + 95x^2 + 53x + 13}{x^6 + 12x^5 + 63x^4 + 184x^3 + 315x^2 + 300x + 125}, \tag{5.88}$$

$$q_1(x) = -\frac{2x^5 + 13x^4 + 40x^3 + 70x^2 + 54x + 9}{x^6 + 12x^5 + 63x^4 + 184x^3 + 315x^2 + 300x + 125}. \tag{5.89}$$

On the other hand, the *ProbA* calculation immediately reveals

$$r_0(X) = -\frac{2}{5}e^{-2X} \qquad r_1(X) = \frac{1}{125}(925X^2 + 50X + 13))e^{-2X}. \tag{5.90}$$

5.2. Let $\Psi(x)$ be the function introduced in the proof of Proposition 3. Use integration by parts to show $\Psi(x)$ has an asymptotic expansion in powers of x^{-1} as $x \to \infty$, and therefore $\Psi(x) \in C^\infty([0, \infty])$. Show also that $\Psi(0) = \frac{1}{2}\pi^{1/2}e^{-i\pi/4}$.

5.3. In Problem I, from the solution of the differential equation for $z_1^{(+)}(\hat{x})$, show that

$$u_1^{(+)}(\hat{x}) = \frac{b_{00}(x_o + \hat{x})}{a_{00}(x_o + \hat{x}) - 1} - \frac{b_{00}(x_o)}{a_{10}(x_o)\hat{x}}, \tag{5.91}$$

$$v_1^{(+)}(\hat{X}) = Z_1^{(+)}(\hat{X}) - Z_1^{(+)}(\infty), \tag{5.92}$$

and

$$Z_1^{(+)}(\infty) = \frac{b_{10}(x_o)}{a_{10}(x_o)} - \frac{b_{00}(x_o)a_{20}(x_o)}{[a_{10}(x_o)]^2}. \tag{5.93}$$

Show also that $u_1^{(-)}(\tilde{x}) = u_1^{(+)}(-\tilde{x})$ and $v_1^{(-)}(\tilde{X}) = v_1^{(+)}(\tilde{X})$.

5.4. Show that

$$e^{ix^2} \int_0^x \Psi(s)e^{-is^2}\,ds = x\Psi(x) + \frac{i}{2}(1 - e^{ix^2}), \tag{5.94}$$

$$e^{ix^2} \int_0^x s^2\Psi(s)e^{-is^2}\,ds = \frac{1}{6}[2x^3\Psi(x) + ix^2 + 1 - e^{ix^2}]. \tag{5.95}$$

and use these results to determine $Z_1^{(+)}(\hat{X})$ from (5.53).

References

1. M. Abramowitz and I. A. Stegun, eds., *Handbook of Mathematical Functions*, National Bureau of Standards, 1964. Reprinted by Dover, New York, 1970.

2. W. Eckhaus, *Asymptotic Analysis of Singular Perturbations*, North-Holland, Amsterdam, 1979.

3. L. E. Fraenkel, On the method of matched asymptotic expansions, Parts I, II and III, *Proc. Cambridge Phil. Soc.*, **65** (1969), 209-231, 233-261 and 263-284.

4. A. M. Il'in, *Matching of Asymptotic Expansions of Solutions of Boundary Value Problems*, Am. Math. Soc., Providence, Rhode Island, 1991.

5. J. Kevorkian and J. D. Cole, *Multiple Scale and Singular Perturbation Methods*, Springer-Verlag, New York, 1996.

6. F. W. J. Olver, *Asymptotics and Special Functions*, Academic Press, New York, 1974.

7. R. E. O'Malley, Jr., *Singular Perturbation Methods for Ordinary Differential Equations*, Springer-Verlag, New York, 1991.

8. L. A. Skinner, Matched asymptotic expansions of integrals, *IMA J. Appl. Math.*, **50** (1993), 77-90.

9. L. A. Skinner, Matched expansion solutions of the first-order turning point problem, *SIAM J. Math. Anal*, **25** (1994), 1402-1411.

10. L. A. Skinner, Asymptotic solution to a class of singularly perturbed Volterra integral equations, *Methods and Applications of Analysis*, **2** (1995), 212-221.

11. L. A. Skinner, Stationary phase theory and passage through resonance, *J. Math. Anal. Appl.*, **205** (1997), 186-196.

12. Donald R. Smith, *Singular-perturbation theory*, Cambridge University Press, Cambridge, 1985.

13. Milton Van Dyke, *Perturbation Methods in Fluid Mechanics*, Parabolic Press, Stanford California, 1975.